ESSENTIALS OF MATH
WITH BUSINESS APPLICATIONS

FIFTH EDITION

C. George Alvey

Professor Emeritus of Business Administration
County College of Morris
Randolph, New Jersey

Marceda Johnson

Kaw Area Technical School
Topeka, Kansas

GLENCOE
McGraw-Hill

New York, New York   Columbus, Ohio   Woodland Hills, California   Peoria, Illinois

Cover Photo: © 1995 Tangent/Nawrocki Stock Photo, Inc. All rights reserved

Calculator Tip photo: Gary Buss/FPG International

Photo credits: page x, Doug Martin; page 2, Eric Futran/Liaison; page 52, Life Images Studios; page 54, Life Images Studios; page 82, Richard Shock/Liaison; page 84, Matt Meadows; page 124, Superstock; page 126, Matt Meadows; page 182, Life Images Studios; page 184, Walter Hodges/Westlight; page 246, Life Images Studios; page 248, KS Studios; page 302, Steven Dunwell/Image Bank; and page 304, Life Images Studios.

Essentials of Math with Business Application, Fifth Edition
Imprint 1997

Send all inquiries to:
Glencoe/McGraw-Hill
936 Eastwind Drive
Westerville, OH 43081

ISBN 0-02-805380-X

Printed in the United States of America.

5 6 7 8 9 10 11 12 13 14 15   066   03 02 01 00 99 98 97

Contents

Unit 3 Decimals: Addition, Subtraction, Multiplication, and Division 83

Unit 4 Working with Fractions 125

Unit 5 Percents 183

Unit 6 Business and Consumer Math 245

Unit 7 Interests and Discounts 303

Preface

The fifth edition of *Essentials of Math with Business Applications* is a comprehensive revision of the *Essentials of Business Math,* fourth edition. While retaining the emphasis on preparing students to be successful in today's workforce and in everyday life, the authors have focused attention on applications in a business environment and on reinforcement by increasing the number of practice exercises and problems. *Essentials of Math with Business Applications* is designed to be user friendly for students and instructors.

Purpose

The goal of this program is to assist students in achieving the Learning Outcomes stated for each Skillbuilder by:

1. Reviewing the fundamentals of addition, subtraction, multiplication, and division through practice with computations involving whole numbers, decimals, fractions, and percents.

2. Relating the students' computational skills to typical business transactions involving interest, discount, payroll, depreciation, retail selling, and checking accounts.

3. Developing students' ability to use shortcut methods and to work with speed and accuracy.

4. Developing students' awareness and usage of the latest business technologies.

This text has been designed so that it can be used by students who are working on an individual basis as well as by students who are being instructed as a group.

Organization

The text is organized in seven units with seven to ten Skillbuilders per unit. Each Skillbuilder states the expected Learning Outcomes and introduces a new mathematical skill using *easy-to-read and easy-to-comprehend* explanations. Skillbuilders contain Math Tips and Calculator Tips. Many also include Business Applications for the skill introduced.

The Learning Outcomes and explanations are followed by practice exercises and application activities. The problems are arranged to progress from the simple to the more complex. Since the ultimate goal is to apply these skills in life and in the workforce, Business Applications are included in most Skillbuilders.

Because of the stated outcomes, the clear and concise explanations and the integration of Business Applications, this text is flexible and can be used in individual or group situations. In addition, the text benefits students at multiple mathematical skill levels. The text has been designed to help students make successful school-to-work transitions.

Major Textbook Changes

- Four-Color Format. A new, four-color format allows functional use of color for emphasis and in the presentation of invoices and other types of business forms.

- Illustrations. This edition uses color illustrations to lend a realistic feel to business application material.

- Skillbuilder Design. The text has a new open design, providing ample space for students to work problems.

- Features. Each Skillbuilder contains a Math Tip and Calculator Tip. Many also include one or more Business Application problems.

- More Problems. This edition includes more problem material to enhance students' mastery of each new skill.

- Competency-Based Skillbuilder. Learning Outcomes introduce each Skillbuilder.

- Self-Check Problems. These problems, which follow a worked-out example, provide immediate reinforcement for the student.

Student Program Components

The student edition of the book is available with optional packaging that includes:

- Numeric Keypad software, which is designed specifically to provide practice and reinforcement of skills needed for the keypad. The student can work at his or her own pace to upgrade accuracy and speed.

- A text-workbook, *Solving Business Problems on the Electronic Calculator,* which provides drills and tests for using the numeric keypad. Now in its fourth edition, this successful book has been used by many instructors to reinforce their students' keypad skills.

Instructor's Support Components

- New to the fifth edition is an *Instructor's Annotated Edition.* Features include:

  - Full student pages with annotations and worked-out solutions shown in place.

  - Correlation of the textbook to the SCANS competencies for easy reference.

- Assignment Guide that provides recommended goals for student progress.

- Cooperative Learning overview of suggestions on how to use cooperative learning activities with this program and how to integrate effective learning strategies in the classroom.

- A new *Instructor's Resource Portfolio* is also available. The folder contains:

  - *Practice Masters,* which provide extra practice for each Skillbuilder

  - *Metric Applications* provides a concise guide to metric units and conversions with problems and exercises for student practice.

  - *Civil Service Test,* a sample of the type of math problems to be expected on typical civil service and employment tests.

  - *Progress Reports,* which can be duplicated for the students to use to mark their progress toward goals.

  - *Thirty Acetate Transparencies* illustrate and summarize selected major concepts.

  - *GED Matrix* correlates the text to topics covered in the GED math test and provides practical suggestions for preparing for the test.

  - *Pretests* are available for each of the seven units. They provide a way for students to ascertain their level of skill and knowledge before beginning a unit.

  - *Unit Tests* can be used to test students' mastery of skills after the completion of each unit.

Acknowledgments

The authors are grateful to the following reviewers, contributors, and business professionals for their contributions during the development of this text: Clare Nelson, Central Piedmont Community College, Charlotte, NC; Doreen E. Stern, Educational Consultant, Yardley, PA; Frank M. Trubiano, Educational Consultant, Long Island City, NY; Sue Schlegel, Consulting Teachers for Business Education, Topeka Public Schools, Topeka, KS: Kathy Markham, Software Application Instructor, Workplace Education Program, Kaw Area Technical School, Topeka, KS; and Arlene Brittain, Real Estate Agent and Retired Reading Specialist, Woodward, OK.

C. George Alvey
Marceda Johnson

U N **1** I T

WHOLE NUMBERS
Addition and Subtraction

Numbers dominate the business world. We use numbers to prepare payrolls, keep track of profit or loss, study industry trends, and so on. This pie chart shows a college's budget for publishing its student newspaper. Using percents, what analysis can you make from the pie chart?

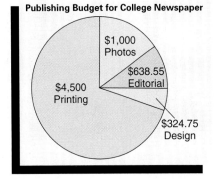

Publishing Budget for College Newspaper

$1,000 Photos
$638.55 Editorial
$4,500 Printing
$324.75 Design

   In this unit we study addition and subtraction of whole numbers and money, including ways to make computation easier and to check our answers.

Chef

Although a chef's main job is to prepare food, a chef needs many of the same qualities as a small business owner. For example, Kim is the head chef at a posh restaurant which also caters parties. Kim manages the kitchen staff, which requires interviewing skills, as well as good interpersonal skills. Kim must also plan a strategy for the preparation of the food, so that each dish is completed on schedule. In addition, Kim must be able to forecast the food requirements; too little food will end with disappointed customers and too much food will result in waste. In order to make the necessary food projections, Kim must be able to estimate, add, and subtract in order to decide on food purchasing amounts.

Math Application

Kim is in charge of planning the "Happy Holidays" dinner for a large insurance company, which means making arrangements for approximately 700 people. The menu for the dinner includes shrimp cocktail and seafood thermidor. The shrimp cocktail requires 350 pounds of shrimp, and the seafood thermidor calls for 225 pounds of shrimp. How much shrimp must the chef buy in order to prepare the dinner?

■ (*a*) Add the necessary amounts to compute the total shrimp required.

Suppose Kim has 110 pounds of shrimp that he purchased and froze last week when it was on sale. How much shrimp must be purchased if the frozen shrimp is used?

■ (*b*) Subtract the number of frozen pounds from the total you found in (*a*).

Critical Thinking Problem

When Kim arrives at the seafood mart to purchase the shrimp, they only have 400 pounds left. What are some alternatives that Kim might use to solve this shortage of shrimp?

SKILLBUILDER 1.1

Reading and Writing Whole Numbers

Learning Outcomes

After completing this skillbuilder, you should be able to:

- Read and write numbers using words.
- Read word names of numbers and write numbers using digits.

Reading and Writing Whole Numbers

Our number system is the **decimal,** or **base 10, system.** It uses the digits 1, 2, 3, 4, 5, 6, 7, 8, 9, and 0. The position of a digit indicates the place value of that digit.

The place value of each digit in a whole number is ten times greater than that of the place to its right. Thus, a digit in the tens place is worth ten times the same digit in the units place, and a digit in the hundreds place is worth ten times the same digit in the tens place.

In numbers greater than three places, commas are used to separate every three digits. Each group of three digits (except the units group) is read as a three-digit number followed by the name of the group to which it belongs.

MATH TIP

Note that a hyphen is used when compound numbers less than 100 are written in words. In other words, all numbers from twenty-one to ninety-nine that are made up of two number names are hyphenated.

EXAMPLE

Write 4,023,905 in words.

SOLUTION

Four million, twenty-three thousand, nine hundred five

 **Self-Check**

1. Write 809 in words.

Self-Check Answer

Eight hundred nine

Problems

Read these numbers, and then write them in words.

1. 83 _____
2. 223 _____
3. 5,908 _____
4. 17,080 _____
5. 32,745 _____
6. 114,811 _____
7. 324,802 _____
8. 1,843,630 _____
9. 916 _____
10. 43 _____
11. 5,492 _____
12. 342,104,007 _____
13. 999 _____
14. 63,389 _____
15. 8,003 _____
16. 11,012 _____
17. 22 _____
18. 157 _____

Writing Whole Numbers Using Digits

We can use a place-value chart to help write numbers using digits. The following chart shows the number thirteen billion, seven hundred fifty million, two hundred forty-five thousand, five hundred seventy-one.

WHOLE NUMBERS														
Billions Group				Millions Group				Thousands Group				Units Group		
Hundred Billions	Ten Billions	Billions	Comma	Hundred Millions	Ten Millions	Millions	Comma	Hundred Thousands	Ten Thousands	Thousands	Comma	Hundred	Tens	Units
	1	2	,	7	5	0	,	2	4	5	,	5	7	1

Problems

Read these numbers, and then write them using digits.

19. Seventy-one   19. _____

20. Sixty-three   20. _____

21. Two hundred forty-nine   21. _____

22. Nine hundred seven   22. _____

23. Sixteen thousand, two hundred three   23. _____

24. One hundred thirty-three thousand, four hundred fifty-nine   24. _____

25. Five million, four hundred fifty thousand, nine   25. _____

26. Three thousand, four hundred seventy-six   26. _____

27. Seven hundred fifty-seven   27. _____

28. Ninety-five   28. _____

29. Eight million, five thousand, nine   29. _____

30. Five hundred nine thousand, nine hundred forty-eight   30. _____

31. Six million, three thousand, eight   31. _____

32. Thirty-nine   32. _____

33. One thousand, four hundred thirty-eight   33. _____

34. Sixty-six   34. _____

35. Four thousand, fifty-three   35. _____

Calculator Tip

Be sure to clear the calculator. Always remember to look for the total symbol on the printout tape or the clear indicator on the screen before beginning any problem.

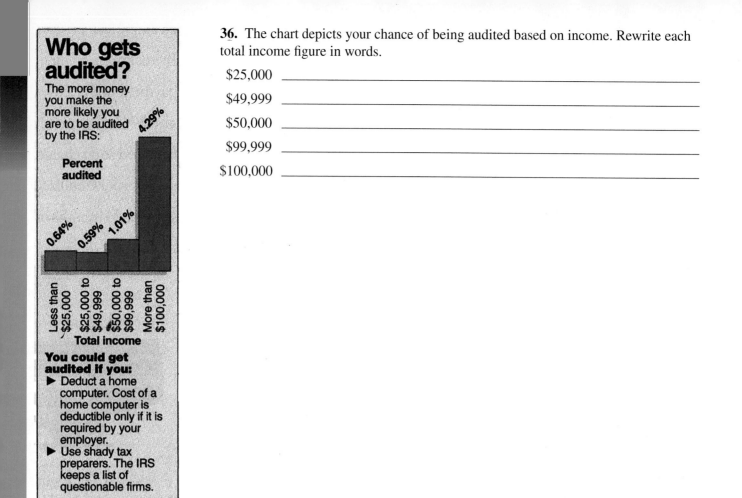

36. The chart depicts your chance of being audited based on income. Rewrite each total income figure in words.

$25,000 _____

$49,999 _____

$50,000 _____

$99,999 _____

$100,000 _____

Student Notes

_____
_____
_____
_____
_____
_____
_____
_____
_____
_____
_____
_____
_____
_____
_____
_____

SKILLBUILDER 1.2

Aligning Digits by Place Value

Learning Outcomes

After completing this skillbuilder, you should be able to:

■ Align a column of numbers according to place value.

Aligning Digits

Because place value is the governing factor in any arithmetic computation, it is essential that numbers be aligned according to the place value of the digits. The units place should be aligned with the units place, the tens place with the tens place, and so on.

MATH TIP

Failure to write numbers clearly and to align them properly is one of the major reasons for errors in arithmetic computations.

EXAMPLE

Align the numbers 300 and 29 so that they can be added.

SOLUTION

Incorrectly Aligned

$$
\begin{array}{r}
300 \\
+29 \\
\hline
590
\end{array}
$$
Incorrect
Answer

Correctly Aligned

$$
\begin{array}{r}
300 \\
+\ 29 \\
\hline
329
\end{array}
$$
Correct
Answer

 ***Self-Check***

Self-Check Answers

Write the following numbers using digits in the columns at the right. Align them correctly both horizontally and vertically.

1. Four hundred three
2. Twenty-five
3. Two
4. Three thousand, twenty-nine
5. Fifty-two

			4	0	3
				2	5
					2
		3	0	2	9
				5	2

Problems

Align the following groups of numbers according to place value. Copy them over so that digits of the same place value are aligned.

1. 472
500
6,843
1,285

1. _____
_____
_____
_____

2. 2,705
63
287
10,540

2. _____
_____
_____
_____

3. 9,453; 9,354; 36; 107

3. _____
_____
_____
_____

4. 10,057; 9,003; 124; 129,646

4. _____
_____
_____
_____

5. The groups of numbers below and on the next page are from different computer printouts. In the first group, the number zero is printed with a slash through it so it will not be confused with the uppercase letter "O." In the second group on the next page, the zero is printed with a dot in its center. Write these groups of numbers in words. Remember that when working with dollars and cents, a decimal point and dollar sign are used. The decimal points are aligned.

a. 66665
61026
35502
8265
95238
54952

5a. _____
_____
_____
_____
_____
_____

Problems

```
b. $500.00   b. _____
    99.37        _____
   105.89        _____
   109.87        _____
   115.68        _____
   123.34        _____
```

6. Write the following group of numbers using digits in a vertical column. Align digits of the same value.

　　Four thousand, fifty-three
　　Six hundred seven
　　Eight thousand, eight hundred eight
　　Seventeen thousand, thirty-two
　　Sixty-four

6. _____
　　　 _____
　　　 _____
　　　 _____
　　　 _____

Write the following numbers using the digits in the columns at the right. Align them correctly both horizontally and vertically.

7. a. One hundred twelve

b. Five thousand, fifty-two

c. Twenty thousand, three hundred one

d. Fifty-six

e. Three hundred two

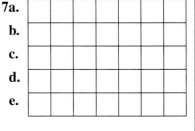

Calculator Tip

To add and subtract money values, set the calculator in the "Add Mode" position if that function is available; if it is not, set the decimal selector to two.

8. a. Two hundred eighty-six

b. Four thousand, five hundred ninety-seven

c. Ninety-nine

d. Nine thousand, two

e. Eight hundred twenty-four

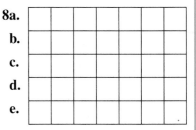

9. a. Seventy-six

b. Four hundred eight-nine

c. Nine hundred ninety-nine

d. Thirty-four

e. eighteen

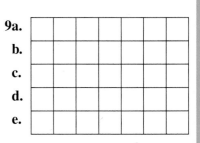

10. **a.** Two million, four hundred sixty-two   **10a.**

    b. Four thousand, fifty-five   **b.**

    c. Seven hundred twenty-five   **c.**

    d. Ninety-two   **d.**

    e. Three hundred sixty-six   **e.**

11. The graphic depicts the average salary of senior executives who graduated from Ivy League universities as $160,620. List the place value of each digit in $160,620.

SKILLBUILDER 1.3
Adding Whole Numbers

After completing this skillbuilder, you should be able to:

■ Add two or more whole numbers.

Adding Whole Numbers

In an addition problem, the numbers being added are called the **addends.** The result of the computation is called the **total,** or **sum.** The order in which the addends are added does not affect the sum.

$$2 + 4 = 6 \qquad 4 + 2 = 6$$

When adding a column of numbers that totals more than 10, it is necessary to carry the first digit of that total to the next column of numbers.

MATH TIP

About one-half of the calculations in business involve addition.

EXAMPLE
Add: 939 + 147 + 265.

SOLUTION

```
 1 2
  939
  147
+ 265
1,351
```

The total of the digits in the units place is 21. The *1* is written under the column that is added, and the *2* is carried to the tens column. The total of the digits in the tens column is 15. The *5* is written under the tens column, and the *1* is carried to the next column to be added, the hundreds column.

 **Self-Check**

Add: 526 + 1,034 + 48 + 885.

Self-Check Answer

_____2,493_____

Problems

Add the numbers mentally. Then write the sums.

	a.	b.	c.	d.	e.	f.	g.	h.	i.	j.
1.	1	3	5	6	2	8	6	9	9	5
	7	9	7	4	6	7	3	9	5	5

	a.	b.	c.	d.	e.	f.	g.	h.	i.	j.
2.	3	6	6	3	1	7	1	2	5	3
	4	6	8	9	1	9	7	5	6	4

	a.	b.	c.	d.	e.	f.	g.	h.	i.	j.
3.	1	6	8	5	2	6	9	5	3	6
	4	2	7	8	3	2	3	6	2	6
	3	5	4	9	7	6	1	9	3	6

	a.	b.	c.	d.	e.	f.	g.	h.	i.	j.
4.	1	7	4	5	6	3	2	4	4	8
	5	4	1	8	5	9	4	3	4	7
	6	4	5	2	7	8	3	6	9	2
	8	2	3	4	4	7	5	8	6	7

Calculator Tip

Answers found with calculators are not always correct. Errors can result from copying numbers incorrectly, striking the wrong key, and so on. If a tape is available, verify the total by checking the original figures against the tape produced by the machine. If a display calculator is used, verify the total by redoing the computations.

Some of the following totals are wrong. Cross out any incorrect answer, and write the correct answer beside it.

	a.	b.	c.	d.	e.	f.	g.	h.	i.	j.
5.	9	7	9	5	3	7	4	9	6	2
	6	5	8	6	7	9	7	6	4	9
	8	2	4	4	9	9	6	7	4	9
	3	6	7	4	3	8	7	3	6	2
	26	20	27	19	22	33	34	25	20	22

	a.	b.	c.	d.	e.	f.	g.	h.	i.	j.
6.	17	10	47	70	53	38	45	85	22	31
	21	25	11	19	42	31	22	30	46	32
	41	34	20	20	41	10	51	53	11	66
	79	57	78	109	136	79	108	168	89	129

Problems

Add the numbers in the following problems.

	a.	b.	c.	d.	e.	f.	g.	h.	i.	j.
7.	89	132	342	45	425	321	1,543	243	2,023	34
	92	88	109	277	231	606	481	1,006	29	930
					456	195	7,328	751	871	256
									530	11

8. Find the total sales for each day of the week.

Representative	M	T	W	T	F
Richards	$ 765	$ 875	$1,254	$ 922	$1,135
Wolf	813	934	983	1,268	820
Bookner	906	1,007	834	785	955
Tobin	1,106	843	912	872	1,010
Wood	751	1,254	840	–0–	975
Total					

9. What is the total amount of gold produced by the United _____
States, South Africa, and Russia (metric tons)?

Golden South Africa
The world produced 2,290 metric
tons of gold in 1993, worth about
$31 billion. Top producers:
Metric tons
620 South Africa
330 U.S.
140 Russia

Source: Bureau of Mines   By Cindy Hall and Suzy Parker, USA TODAY

Student Notes

_____
_____
_____
_____
_____
_____
_____
_____
_____
_____
_____
_____
_____
_____
_____
_____
_____
_____
_____

SKILLBUILDER 1.4

Developing Speed in Addition

Learning Outcomes

After completing this skillbuilder, you should be able to:

- Add whole numbers quickly by adding tens and then ones.
- Add whole numbers quickly by grouping sums of 10.

Add Tens First

One method of adding two-digit numbers quickly is mentally to add all the tens digits and one of the units digits first. Then add the remaining units digits to the first total.

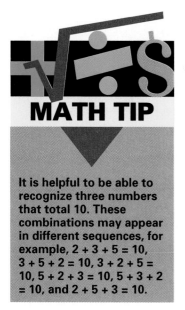

MATH TIP

It is helpful to be able to recognize three numbers that total 10. These combinations may appear in different sequences, for example, 2 + 3 + 5 = 10, 3 + 5 + 2 = 10, 3 + 2 + 5 = 10, 5 + 2 + 3 = 10, 5 + 3 + 2 = 10, and 2 + 5 + 3 = 10.

EXAMPLE

Add: 13 + 49 + 52.

SOLUTION

To add 13 + 49 + 52, first add 13 + 40 + 50 = 103. Then add 103 + 9 = 112 and 112 + 2 = 114.

$$
\begin{array}{ll}
13 & 13 \\
49 & 40 \\
+\,52 & +\,50 \\
\end{array}
$$

103 + 9 = 112

112 + 2 = 114

▶ **Self-Check**

Use the method described here to find 55 + 83.

Self-Check Answer

_____138_____

Problems

Find the sum in each of the following problems Use the method described here.

	a.	b.	c.	d.	e.	f.	g.	h.	i.	j.
1.	26	19	43	79	68	54	99	77	93	84
	57	86	76	44	87	92	32	29	61	39

	a.	b.	c.	d.	e.	f.	g.	h.	i.	j.
2.	22	38	57	49	83	19	97	86	46	28
	16	41	35	86	71	99	31	42	59	57
	83	96	48	28	68	67	46	49	32	78

	a.	b.	c.	d.	e.	f.	g.	h.	i.	j.
3.	49	42	19	39	68	21	73	59	84	37
	56	61	93	67	86	56	78	67	47	28
	62	98	42	85	40	69	84	82	10	55
	78	48	78	67	69	38	14	66	87	75

	a.	b.	c.	d.	e.	f.	g.	h.	i.	j.
4.	75	48	62	41	31	86	56	83	77	81
	15	37	28	60	48	53	20	95	17	62
	53	82	74	39	99	29	33	51	36	42
	98	52	85	58	57	39	68	59	93	57
	84	99	97	75	89	73	99	33	51	29

Calculator Tip

To add whole numbers, set the decimal-point selector to zero.

Add By Grouping

Another way to develop speed in addition is to add using groups of two or more digits that total 10.

$$
\begin{array}{ccccccccc}
1 & 2 & 3 & 4 & 5 & 6 & 7 & 8 & 9 \\
+9 & +8 & +7 & +6 & +5 & +4 & +3 & +2 & +1 \\
\hline
10 & 10 & 10 & 10 & 10 & 10 & 10 & 10 & 10
\end{array}
$$

First add any digits in a column that total 10. Then add this total to the remaining digits.

EXAMPLE

Add: 96 + 54 + 63 + 12 + 55.

SOLUTION

The digits connected by lines total 10.

$$10 = \begin{cases} \overset{2}{96} \\ 54 \end{cases} = 10$$
$$10 = \begin{cases} 63 \\ 12 \\ +55 \end{cases} = 10$$

280

Note that 10 tens + 10 tens + 6 tens = 26 tens, or 260. Then 260 + 10 + 10 = 280.

▶ **Self-Check**

Add by using groups that total 10:
27 + 69 + 71 + 43 + 81.

Self-Check Answer

291

Problems

Add the numbers in these problems by using groups that total 10.

	a.	b.	c.	d.	e.	f.	g.
5.	216	547	813	5,426	4,875	498	$928
	461	164	726	1,251	9,234	953	163
	628	432	652	4,943	1,768	691	42
	593	634	295	3,942	3,312	462	5
	836	522	134	5,627	6,241	312	675
						249	12
						527	415

6.	$9.28	3,613	156	492	333	611	738
	2.86	9,149	904	816	777	994	327
	1.59	2,572	106	329	888	653	673
	4.62	8,426	936	817	222	457	437
	5.43	6,321					
	3.52	4,157					
	6.21	2,212					

Using either method described, find the sum of each of the following groups.

	a.	**b.**	**c.**	**d.**	**e.**
7.	427	506	468	683	492
	68	82	49	91	84
	127	104	106	121	114
	653	602	608	801	590
	371	360	346	294	361

	f.	**g.**	**h.**	**i.**	**j.**
	92	955	86	365	513
	742	91	256	49	47
	436	484	427	559	562
	328	379	258	382	432
	685	764	513	365	607

8. Without using a calculator, compute the total number of espionage cases that were uncovered between 1950 and 1989.

Answer
_____

'80s: The spying decade
Number of espionage cases uncovered in the USA, by decade of the Cold War:

48
24
19
12

1950–'59 1960–'69 1970–'79 1980–'89
Source: Department of Defense

By Anne R. Carey and Sam Ward, USA TODAY

Student Notes

_____
_____
_____
_____
_____
_____
_____
_____
_____

SKILLBUILDER 1.5

Checking Addition

After completing this skillbuilder, you should be able to:

■ Check the accuracy of addition.

Checking Addition

Never assume that the addition of a column of figures is correct. If figures were copied from another source, check the copy against the original to be sure that the figures were copied accurately. Check the accuracy of addition done by hand by adding the same figures in the reverse order.

MATH TIP

Remember, the possibility of error is introduced when copying numbers from one document to another.

EXAMPLE

Add and check: $385 + 54 + 915 + 446$.

SOLUTION

1,800

Then add up.
$$\begin{array}{r} 385 \\ 54 \\ 915 \\ 446 \\ \hline 1,800 \end{array}$$
First add down.

▶ **Self-Check**

Add and check: $1,024 + 378 + 2,592 + 74$.

Self-Check Answer

4,068

Problems

Copy each of the following groups of figures in the space provided, find the total, and then check the total by adding in reverse order. Try using the speed methods presented in Skillbuilder 1.4.

1.	**2.**	**3.**	**4.**
57	386	$6,321	$54,943
36	159	246	7,156
48	514	2,179	63,041
24	212	1,532	6,119

5.	**6.**	**7.**	**8.**
35	457	$15,726	$18,182
16	123	3,168	717
84	316	9,003	4,111
25	286	213	13,155
57	934	1,124	2,435

9.	**10.**	**11.**	**12.**
10,345	6,098	876	$12,387
5,876	23,086	12,432	4,592
11,490	10,137	2,376	5,231
30,467	9,437	690	10,429
3,041	897	4,639	234

13. This record shows the number of bicycles sold during Bike World's sale week. How many were sold each day?

14. Complete the weekly production report. How many units were produced in one week by each employee?

DAILY SALES—UNITS

Bicycle	M	T	W	TH	F
28″ mountain bike	5	8	6	–0–	9
10-speed	11	10	14	12	15
12-speed	15	16	18	17	24
20″ BMX	4	–0–	3	3	5
12″ RMX	2	2	0	1	7
Total					

WEEKLY PRODUCTION REPORT

Employee Number	Number of Units					Total Units
	M	T	W	TH	F	
31045	65	62	67	61	64	____
10641	63	58	60	61	59	____
42352	54	53	60	59	58	____
10914	68	74	70	76	71	____
91837	59	58	60	60	61	____

Problems

15. Determine how much was spent each month for utilities. Double-check your addition by adding in the reverse order. Remember to align the decimal points and include a dollar sign in your answer.

Item	January	February	March	April	May	June
			UTILITY BILLS			
Gas	$ 92.53	$ 102.10	$ 73.90	$ 54.35	$ 44.11	$ 43.08
Electric	78.14	77.56	74.51	93.04	113.25	137.34
Water	–0–	58.97	–0–	–0–	87.43	–0–
Telephone	26.80	55.20	62.15	42.32	74.10	51.13
Total	_____	_____	_____	_____	_____	_____

16. Find the total sales for each day shown on the sales report. Check your answers by adding in the reverse order.

Item	May 23	May 24	May 25	May 26	May 27
		WEEKLY SALES REPORT			
Decals	$ 58.00	$ 61.00	$ 43.00	$ 49.00	$ 36.00
Pennants	63.00	70.00	54.25	57.75	49.00
T-shirts	69.00	57.00	63.00	42.00	39.00
Coffee mugs	55.50	45.00	37.50	42.00	15.00
Steins	41.25	46.75	30.25	33.00	33.00
Total	_____	_____	_____	_____	_____

Calculator Tip

When entering money amounts, the results will be more accurate if you mentally read the number as "three, four, point, two, three" instead of "thirty-four dollars and twenty-three cents."

Problems

17. Complete the production report. Check your totals by adding in the reverse order.

UNIT PRODUCTION REPORT				
Name	**July 1–7**	**July 8–14**	**July 15–21**	**July 22–28**
Sanchez	760	1,212	836	975
Mlynek	685	790	1,005	930
Simpson	725	935	824	1,207
Sumpter	760	1,230	980	874
Woods	770	1,120	915	1,245
Total	____	____	____	____

18. What was the total amount of the payroll for each of the weeks shown?

NET PAYROLL REPORT				
Name	**July 1–7**	**July 8–14**	**July 15–21**	**July 22–28**
Sanchez	$280.20	$446.84	$308.22	$359.46
Mlynek	252.55	291.26	370.52	342.87
Simpson	267.29	344.72	303.79	445.00
Sumpter	280.20	453.48	361.31	322.23
Woods	283.88	412.92	337.34	459.01
Total	____	____	____	____

19. Have all 57 of the nations been graphed in the art below?

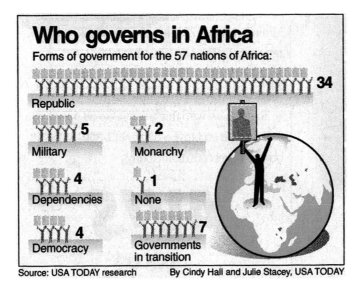

Who governs in Africa
Forms of government for the 57 nations of Africa:

Republic 34
Military 5
Monarchy 2
Dependencies 4
None 1
Democracy 4
Governments in transition 7

Source: USA TODAY research   By Cindy Hall and Julie Stacey, USA TODAY

NAME: _____

DATE: _____

SKILLBUILDER 1.6

Adding Horizontally and Vertically

Learning Outcomes

After completing this skillbuilder, you should be able to:

- Add whole numbers horizontally.
- Check addition by adding horizontally and vertically.

Horizontal Addition

Figures written in a horizontal line rather than in a column can be rewritten in a vertical column before adding, but this is not always necessary.

Horizontal addition is performed in the same way as vertical addition. Each digit is added to its counterpart in each number: units are added to units, tens to tens, hundreds to hundreds, and so on.

MATH TIP

When adding horizontally, be careful to match the proper digits. Horizontal addition can be checked by adding the numbers in the reverse order.

EXAMPLE

Add horizontally: $157 + 316 + 942$.

SOLUTION

First add the units digits. Add the *1* (the number carried) to the tens digits. Add the *1* (the number carried) to the hundreds digits.

$$157 + 316 + 942 = 1,415$$
$$7 + 6 + 2 = 15 \text{ units}$$
$$1 + 5 + 1 + 4 = 11 \text{ tens}$$
$$1 + 1 + 3 + 9 = 14 \text{ hundreds}$$

14 hundreds + 1 ten + 5 units = 1,415

 **Self-Check**

Add the numbers from left to right. Check by adding from right to left.

$23 + 56 + 37 + 81 + 19$

Self-Check Answer

216

Problems

Add the numbers in these problems from left to right. Check your totals by adding from right to left.

1. 6 + 3 + 4 + 9 + 2   1. _____

2. 57 + 19 + 82 + 36 + 96   2. _____

3. 5 + 8 + 1 + 7 + 5   3. _____

4. 54 + 22 + 10 + 83 + 44   4. _____

5. 7 + 4 + 6 + 3 + 9   5. _____

6. 99 + 14 + 35 + 23 + 36   6. _____

7. 8 + 4 + 5 + 6 + 2   7. _____

8. 139 + 643 + 435 + 815   8. _____

Horizontal and Vertical Addition

Many business forms require numbers to be added both horizontally and vertically. For example, the following problems are excerpts from ledgers. The total of all the vertical columns should equal the total of all the horizontal rows. This procedure, called **crossfooting,** is the most positive check for accuracy.

EXAMPLE

Add the numbers horizontally and vertically. Then add the totals of the vertical columns and horizontal rows.

234	198	3,024	
2,519	335	488	
307	882	1,360	

SOLUTION

234	198	3,024	3,456
2,519	335	488	3,342
307	882	1,360	2,549
3,060	1,415	4,872	9,347

Problems

5	4	3	9	21
007	8	1	2	18
004	7	9	3	23
005	1	6	8	20
021	20	19	22	82

Add the numbers horizontally and vertically. Then add the totals of the vertical columns and horizontal rows.

9.

16	89	57	
45	65	55	
93	22	18	
44	39	56	

10.

1,526	142	901	
803	345	541	
2,000	93	488	
826	329	281	

11.

159	345	80	
1,345	876	1,250	
947	3,222	435	
500	1,303	612	

12. On the following sales report, find: the daily total of all the sales; the total for the week for each sales representative; and the grand total for the week.

Sales Representative	SU	M	T	W	TH	F	SA	TOTAL
Ling	$ 2,409	$–0–	$2,712	$3,930	$ –0–	$2,664	$2,670	_____
Wolf	–0–	3,750	–0–	2,625	2,790	2,889	3,450	_____
Bookner	2,727	3,690	–0–	–0–	3,450	–0–	2,535	_____
Ameredes	2,295	–0–	3,855	2,616	–0–	2,958	2,796	_____
Perez	–0–	2,631	–0–	2,895	2,205	3,600	2,682	_____
Total	_____	_____	_____	_____	_____	_____	_____	_____

WEEKLY SALES REPORT

Problems

13. Find the total number of hours worked each month. Then find the 6-month total for each worker, as well as the total hours worked by all the workers in the 6-month period.

						HOURS WORKED	
Worker	January	February	March	April	May	June	Total
Arnold	168	152	176	168	168	176	____
Bernard	173	160	176	168	168	176	____
Einhorn	168	160	176	173	168	186	____
Grand	168	165	181	168	168	176	____
Johnson	168	165	186	178	168	181	____
Singer	168	160	168	164	168	176	____
Victor	168	160	186	168	168	181	____
Total	____	____					____

15. If Greg Norman wins one more tournament, what will his total earnings equal.

Answer

_____

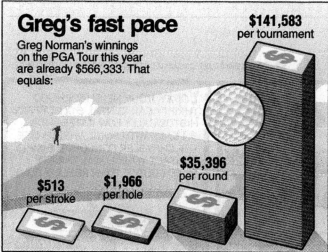

Greg's fast pace

Greg Norman's winnings on the PGA Tour this year are already $566,333. That equals:

$141,583 per tournament

$35,396 per round

$1,966 per hole

$513 per stroke

Source: USA TODAY research   By Chris Fruitrich and Stephen Conley, USA TODAY

SKILLBUILDER 1.7

Increasing Speed in Adding Large Groups of Numbers

Learning Outcomes

After completing this skillbuilder, you should be able to:

■ Use subtotals to increase speed in addition.
■ Use the accountant's method of addition to add large groups of numbers.

Use Smaller Groups of Numbers

One way to add large groups of numbers is to divide the numbers into smaller groups. That is, break a large group of numbers into two or more subgroups and add each of these subgroups. Then add the totals of the subgroups to find the grand total.

MATH TIP

To improve speed and accuracy in addition, always make sure that numbers are written clearly and in uniform, straight columns and lines.

EXAMPLE

Find the total of 428, 645, 27, 156, 42, 583, 897, 273, 6, and 913.

SOLUTION

Write the numbers in a column and break them into two subgroups.

```
428
645
 27
156
 42      a. _____1,298_____
583
897
273
  6
913      b. _____2,672_____
Total    c. _____3,970_____
```

Self-Check

Use subgroups to add: 563 + 294 + 18 + 761 + 109 + 386 + 74 + 3,602 + 866 + 413.

Self-Check Answer

_____7,086_____

Problems

Add the numbers in these problems by finding subtotals at the places indicated and then by finding the overall total.

1. 1,687
   765
    45
 4,525
 <u>8,441</u>
  893
  757
 5,520
   43
 <u>3,192</u>

1. a. _____

b. _____

c. _____

2. 3,786
 4,508
 1,209
 7,459
 <u>6,663</u>
 2,092
 5,698
 4,983
 3,678
 <u>1,954</u>

2. a. _____

b. _____

c. _____

3. 15,672
  8,534
  1,345
 87,354
 <u>30,496</u>
  4,598
 15,927
 48,453
  4,899
 <u>41,746</u>

3. a. _____

b. _____

c. _____

4. 609,576
  198,075
 1,246,934
    4,508
  <u>246,055</u>
   42,233
  120,054
     438
   46,196
  <u>808,417</u>

4. a. _____

b. _____

c. _____

Problems

5.   555,050
      1,246,583
         52,176
        459,021
      177,749
      ─────────
      2,243,265
        568,000
          3,491
         44,400
        487,237
      ─────────

5. a. _____

b. _____

c. _____

Use the Accountant's Method

Another way to break down large groups of numbers is the **accountant's method of addition.** Each column of digits is added separately, and then the column totals are added.

EXAMPLE

Use the accountant's method to add: 4,629 + 816 + 3,937 + 1,014.

SOLUTION

Add the units first (9 + 6 + 7 + 4 = 26), the tens next (2 + 1 + 3 + 1 = 7), the hundreds next (6 + 8 + 9 + 0 = 23), and then the thousands (4 + 3 + 1 = 8). Finally, add the column totals.

$$
\begin{array}{r}
4,629 \\
816 \\
3,937 \\
+1,014 \\
\hline
26 \\
07 \\
2\,3 \\
8 \\
\hline
10,396
\end{array}
$$

Note that the column totals are staggered. When a column total is less than 10, put a zero in front of the number to ensure that the columns are aligned properly. This method of addition is particularly useful in work that is frequently interrupted.

▶ ***Self-Check***

Use the accountant's method to add: 503 + 738 + 4,216 + 59 + 870 + 1,518.

Self-Check Answer

7,904

Problems

Add the numbers in these problems using the accountant's method of addition.

6.   657
  2,097
     43
  4,045
   293
  1,156

Answers

6. _____

7. 5,090
  1,534
    54
   441
  2,000
  4,080

7. _____

8. 10,921
  5,032
   545
  22,318
  4,112
  1,401

8. _____

9.   602
  34,839
  3,034
  6,721
    19
  3,501

9. _____

Problems

5.   555,050
1,246,583
52,176
459,021
<u>177,749</u>
2,243,265
568,000
3,491
44,400
<u>487,237</u>

5. a. _____

b. _____

c. _____

Use the Accountant's Method

Another way to break down large groups of numbers is the **accountant's method of addition.** Each column of digits is added separately, and then the column totals are added.

EXAMPLE

Use the accountant's method to add: $4{,}629 + 816 + 3{,}937 + 1{,}014$.

SOLUTION

Add the units first ($9 + 6 + 7 + 4 = 26$), the tens next ($2 + 1 + 3 + 1 = 7$), the hundreds next ($6 + 8 + 9 + 0 = 23$), and then the thousands ($4 + 3 + 1 = 8$). Finally, add the column totals.

$$
\begin{array}{r}
4{,}629 \\
816 \\
3{,}937 \\
+1{,}014 \\
\hline
26 \\
07 \\
2\ 3 \\
8 \\
\hline
10{,}396
\end{array}
$$

Note that the column totals are staggered. When a column total is less than 10, put a zero in front of the number to ensure that the columns are aligned properly. This method of addition is particularly useful in work that is frequently interrupted.

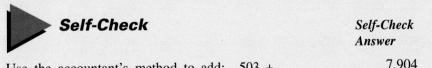

Self-Check

Use the accountant's method to add: $503 + 738 + 4{,}216 + 59 + 870 + 1{,}518$.

Self-Check Answer

<u>7,904</u>

Problems

Add the numbers in these problems using the accountant's method of addition.

6.   657
   2,097
     43
   4,045
    293
   1,156

6. _____

7. 5,090
   1,534
     54
    441
   2,000
   4,080

7. _____

8. 10,921
   5,032
    545
  22,318
   4,112
   1,401

8. _____

9.   602
  34,839
   3,034
   6,721
    19
   3,501

9. _____

Problems

10. On the following weekly report, find:
   a. The weekly total for each department for each shift.
   b. The day total of all sales for the day for each shift.
   c. The weekly total for each shift.
   d. The day total of all sales for the week.
   Note: The total of the Day Total row should equal the total of the Weekly Total column.

WEEKLY SALES REPORT

Department	SU	M	T	W	TH	F	SA	Weekly Total
Clothing	$453	$395	$404	$328	$420	$450	$615	_____
Pharmacy	320	374	291	298	312	345	380	_____
Hardware	95	114	105	83	118	154	165	_____
Notions	192	183	156	133	94	190	125	_____
Sports	108	95	82	211	106	174	115	_____
Shift 1 Total	____	____	____	____	____	____	____	_____
Clothing	729	1,105	865	1,346	1,428	1,515	1,862	_____
Pharmacy	540	495	611	502	1,003	908	1,432	_____
Hardware	1,230	765	840	903	745	1,762	2,145	_____
Notions	659	583	490	553	429	593	805	_____
Sports	730	1,125	980	1,358	949	1,458	1,652	_____
Shift 2 Total	____	____	____	____	____	____	____	_____
Clothing	564	435	446	358	475	497	712	_____
Pharmacy	482	654	593	605	983	853	727	_____
Hardware	1,154	839	1,157	720	835	1,156	1,341	_____
Notions	634	602	545	524	487	627	790	_____
Sports	874	972	1,382	892	869	821	945	_____
Shift 3 Total	____	____	____	____	____	____	____	_____
Day Total	____	____	____	____	____	____	____	_____

Calculator Tip

If a calculator has a repeat entry feature, once a number has been entered, the number can be repeated simply by operating the plus key.

11. Use the accountant's method to compute the total number of assists for these players.

Answers

_____

Closing in on record

Utah's John Stockton — who has 1,003 assists this season — could break the NBA career record for assists next season. The leaders:

RECORD FOR ASSISTS

9,921	9,887	9,355	9,061

Magic Johnson Oscar Robertson John Stockton Isiah Thomas

Source: NBA

By John Riley and Marcy E. Mullins, USA TODAY

Student Notes

_____

_____

_____

_____

_____

_____

_____

_____

_____

_____

_____

_____

_____

_____

_____

_____

_____

SKILLBUILDER 1.8

Rounding Numbers and Estimating Answers

Learning Outcomes

After completing this skillbuilder, you should be able to:

- Round whole numbers to a specified place.
- Estimate a sum using rounded numbers.

Rounding Numbers

Rounding is replacing an actual figure or amount with an approximate figure. For example, sales of 19,883 items could be rounded to sales of around 20,000 items for a verbal report. Money amounts are often rounded to the nearest cent. A sales tax of $0.686 is rounded to $0.69.

To round a number, first locate the place to which the number will be rounded. Then look at the digit to the right of this place. If this digit is 5 or greater, increase the digit in the rounding place by 1. If this digit is less than 5, leave the digit in the rounding place as is. Digits to the right of the rounding place are replaced by zeros.

MATH TIP

Estimating will let you know if your answer is reasonable. If you want to know if your estimate is correct, check by adding the numbers in a different order.

EXAMPLE

Round 5,627 to the tens place.

SOLUTION

Put a line under the digit in the tens place. Then place an arrow over the digit to the right.

$$\downarrow$$
$$5,627$$

Since 7 is greater than 5, increase by 1 the digit in the tens place. The tens digit becomes 3 and the units digit is replaced with a zero.

5,627 rounds to 5,630

 **Self-Check**

Round 5,627 to the hundreds place.

Self-Check Answer

5,600

Problems

Round each number as indicated.

1. 553 to the nearest ten

2. 8,605 to the nearest hundred

3. 32,609 to the nearest ten

4. 52,999 to the nearest thousand

5. 17,839 to the nearest ten

6. 37,056 to the nearest thousand

7. 2,392 to the nearest hundred

8. 5,949 to the nearest ten

9. 763 to the nearest hundred

10. 547 to the nearest thousand

11. 16,572 to the nearest hundred

12. 46,219 to the nearest ten thousand

13. 5,449 to the nearest ten

14. 237,219 to the nearest thousand

15. 776 to the nearest ten

16. 24,986 to the nearest hundred

17. 17,605 to the nearest ten thousand

18. 4,762 to the nearest thousand

1. _____

2. _____

3. _____

4. _____

5. _____

6. _____

7. _____

8. _____

9. _____

10. _____

11. _____

12. _____

13. _____

14. _____

15. _____

16. _____

17. _____

18. _____

Estimating Answers in Addition

Estimating answers helps to uncover obvious errors in computations. Being able to estimate answers is especially important when using electronic calculators.

One way to estimate an answer to an addition problem is to round each number to its largest place. Then add the rounded numbers. The estimate will be close to the actual answer.

EXAMPLE

Estimate the sum: $322 + 576 + 349 + 105$. Then determine the actual sum.

SOLUTION

Estimate	*Actual*
300	322
600	576
300	349
100	105
1,300	1,352

Problems

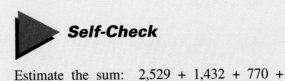

Self-Check

Estimate the sum: 2,529 + 1,432 + 770 + 4,200. Then determine the actual sum.

Self-Check Answer

Estimate <u>8,800</u>

Actual <u>8,931</u>

Estimate the answer to each problem by rounding each number to its largest place and adding the estimates. Then add the numbers to determine the actual totals. Compare your estimated answer with your actual answer to see if your actual answer is reasonable.

19.		**20.**		**21.**	
Actual	*Estimate*	*Actual*	*Estimate*	*Actual*	*Estimate*
37		88		47	
61		8		62	
43		52		19	
89		96		39	
52		12		65	
46		68		22	
—		—		—	

22.		**23.**		**24.**	
Actual	*Estimate*	*Actual*	*Estimate*	*Actual*	*Estimate*
737		209		785	
242		454		362	
495		298		411	
315		549		530	
389		897		647	
669		242		645	
—		—		—	

Calculator Tip

Estimate addition before you enter numbers into the calculator to assist in recognizing errors.

25.		**26.**		**27.**	
Actual	*Estimate*	*Actual*	*Estimate*	*Actual*	*Estimate*
3,067		4,381		7,428	
4,198		7,007		3,664	
4,751		3,436		1,283	
2,009		3,605		4,245	
4,113		1,651		6,495	
4,851		1,770		2,629	
—		—		—	

Problems

28. What is the total number of injuries per 100,000 participants in these sports? Estimate the answer and then calculate the exact answer.

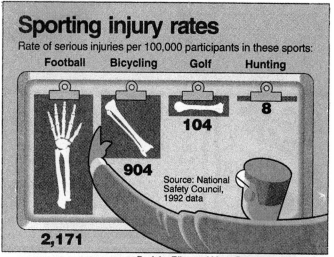

Sporting injury rates

Rate of serious injuries per 100,000 participants in these sports:

Football   Bicycling   Golf   Hunting

104

8

904

Source: National Safety Council, 1992 data

2,171

By John Riley and Marty Baumann, USA TODAY

Student Notes

SKILLBUILDER 1.9

Subtracting Numbers

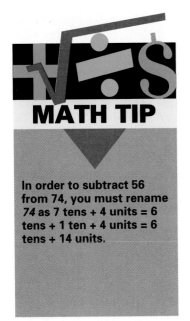

Learning Outcomes

After completing this skillbuilder, you should be able to:

- Find the difference between two numbers.
- Check subtraction by using addition.

Subtracting Whole Numbers

In subtraction, each digit in the **subtrahend** is subtracted from the digit in the same place (that is, having the same place value) in the **minuend.** The result of the subtraction is called the **difference.** If a digit in the subtrahend is larger than the digit above it, it is necessary to borrow before subtracting.

MATH TIP

In order to subtract 56 from 74, you must rename **74** as 7 tens + 4 units = 6 tens + 1 ten + 4 units = 6 tens + 14 units.

EXAMPLE

Subtract: 74 – 56.

SOLUTION

Ten units are borrowed from the tens column and added to the units column.

$$\begin{array}{r} {}^{6\ 14} \\ 7\!\!\!/4\!\!\!/ \quad \text{Minuend} \\ -56 \quad \text{Subtrahend} \\ \hline 18 \quad \text{Difference} \end{array}$$

▶ *Self-Check*

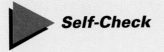

Subtract: 325 – 158.

Self-Check Answer

_____167_____

Problems

Find the difference in each of these problems.

	a.	b.	c.	d.	e.	f.	g.
1.	97	42	40	61	73	89	65
	44	28	27	39	59	24	25

	a.	b.	c.	d.	e.	f.	g.
2.	561	402	852	820	818	211	604
	429	128	749	427	239	106	355

Checking Subtraction

Check subtraction by adding the subtrahend to the difference; the sum should equal the minuend.

EXAMPLE

Subtract 4,580 from 7,870. Check your answer.

SOLUTION

Check

```
    717
  7,870   Minuend        4,580   Subtrahend
 −4,580   Subtrahend   + 3,290   Difference
  3,290   Difference      7,870   Minuend
```

 **Self-Check**

Subtract: 1,398 − 799. Check your answer.

Self-Check Answer

_____599_____

Problems

Find the difference in each problem. Check by adding the subtrahend to the difference.

	a.	b.	c.	d.	e.
3.	1,807 643	436 278	9,002 4,928	954 643	6,598 2,693
4.	15,087 13,258	47,931 28,853	356,803 58,999	134,628 99,999	67,550 32,536
5.	3,907 − 948	810 − 280	9.407 −1,712	7,236 − 289	5,933 −3,721
6.	8,079 − 555	423 − 294	679 −194	25,003 −2,735	86,004 −9,627
7.	8,202 −3,917	7,206 −1,624	43,748 − 2,367	8,237 − 62	27,034 −8,237
8.	80,529 − 21,076	12,701 −9,514	3,933 − 387	95,442 − 86,857	6,465 −2,975
9.	3,050,092 −1,552,325	8,000,235 −1,230,768	7,826,532 −3,491,673	2,425,324 −342,259	

Problems

Calculator Tip

To subtract whole numbers, set the decimal selector to zero.

10. Net sales is the difference between sales and sales returns. Determine the net sales for each day of the week for Essex Street Market.

ESSEX STREET MARKET						
	SU	M	T	W	TH	F
Sales	$10,943	$8,678	$7,586	$8,840	$9,624	$12,955
Sales returns	217	642	258	225	195	381
Net sales						

11. What is the difference in the amount spent between the second- and third-largest spenders?

_____

TV ad giants

Companies that spent most to advertise on network TV programs in 1993:

(in millions)

Procter & Gamble $624

General Motors $454

Philip Morris $374

PepsiCo $310

Source: Television Bureau of Advertising

By Marcia Staimer, USA TODAY

SKILLBUILDER 1.10

Subtracting Horizontally and Estimating Answers

Learning Outcomes

After completing this skillbuilder, you should be able to:

- Subtract vertically and horizontally.
- Estimate answers in subtracting using rounded numbers.

Subtracting Horizontally

In horizontal subtraction, each digit in the subtrahend is subtracted from the digit in the same place in the minuend, units from units, tens from tens, and so on.

Subtraction can also be checked by computing both horizontally and vertically. The total of the horizontal minuends *less* the total of the horizontal subtrahends should equal the total of the horizontal differences.

MATH TIP

The decimal point is understood in a whole number; therefore, the decimal point does not have to be written. We must, however, remember the decimal point is to the right of the number when solving problems.

EXAMPLE

Subtract horizontally. Then find vertical column totals and subtract.

$$15 - 5$$
$$20 - 3$$
$$56 - 4$$

SOLUTION

$$15 - 5 = 10 \qquad 15 - 5 = 10$$
$$20 - 3 = 17 \qquad 20 - 3 = 17$$
$$56 - 4 = 52 \qquad \underline{56 - 4 = 52}$$
$$ \overline{91 - 12 = 79}$$

Problems

Complete the following records and forms by subtracting horizontally.

1.

Item	Beginning	End	Sales
28″ mountain bike	40	12	—
10-speed	90	28	—
12-speed	125	35	—
20″ BMX	45	30	—
12″ RMX	35	23	—

2.

Item	Beginning	End	Sales
Blouses	398	98	—
Dresses	435	124	—
Jackets	145	53	—
Jeans	503	218	—
Shirts	258	58	—
Shorts, Boys'	382	151	—
Shorts, Girls'	312	144	—
Skirts	482	224	—
Socks	505	132	—
Sweaters	165	76	—

3.

Account Number	Billed	Discount Allowed	Received
115	$467.98	$18.72	_____
125	190.23	3.80	_____
155	102.45	3.07	_____
165	93.25	0.93	_____
185	303.46	9.10	_____
195	88.72	0.89	_____

4.

Account Number	Amount Due	Amount Paid	Balance Due
230	$1,456	$ 956	—
240	890	250	—
250	2,312	1,312	—
260	563	250	—
270	934	450	—
280	450	225	—

Problems

5. Ernest Enterprises maintains a check register in which the company records all checks that it writes to pay for invoices for the merchandise it has purchased. The amount of the check is the amount of the invoice less any cash discount. Complete the check register and check your computations.

CHECK REGISTER
ERNEST ENTERPRISES
JUNE 17, 19—

Check Number	Amount of Invoice	Cash Discount	Amount of Check
577	$ 357.63	$ 10.73	_____
578	1,982.75	99.14	_____
579	145.90	2.92	_____
580	55.67	–0–	_____
581	598.50	29.93	_____
582	346.38	10.39	_____
583	28.25	–0–	_____
584	481.09	19.24	_____
Total	_____	_____	_____

6. Complete the inventory and check your computations.

BIKE WORLD
INVENTORY FOR WEEK ENDING APRIL 28, 19—

Item	Value of Beginning Inventory	Amount of Sales	Value of Inventory on Hand
28" mountain bike	$3,399.60	$2,379.72	_____
10-speed	8,099.10	5,579.38	_____
12-speed	12,811.25	9,224.10	_____
20" BMX	3,143.25	1,047.75	_____
12" RMX	1,748.95	599.64	_____
Total	_____	_____	_____

Problems

7. Complete the inventory and check your computations.

SANDY'S SPORTSWEAR INC.
INVENTORY FOR WEEK ENDING JULY 11, 19—

Item	Value of Beginning Inventory	Amount of Sales	Value of Inventory on Hand
Blouses	$2,358.09	$954.65	_____
Tank Tops	987.87	254.98	_____
Jeans	3,459.22	1,040.75	_____
Shoes, Canvas	1,009.45	550.34	_____
Shoes, Leather	2,104.50	973.62	_____
Shorts	1,985.02	654.82	_____
Skirts	3,987.38	1,269.02	_____
Tennis dresses	1,753.72	703.45	_____
Total	_____	_____	_____

8. The following charts show the number of business starts and business failures for the months Jan. to June and July to Dec. in three regional areas. For each region, determine the difference in business starts and failures.

Business Starts	Jan. to June	July to Dec.	Difference
Region A	4,964	5,455	_____
Region B	19,877	21,708	_____
Region C	39,966	41,948	_____
Total	_____	_____	_____

Business Failures	Jan. to June	July to Dec.	Difference
Region A	463	357	_____
Region B	1,733	1,514	_____
Region C	4,070	3,998	_____
Total	_____	_____	_____

Problems

Estimating Answers in Subtraction

MATH TIP

Remember, an estimate will not let you check the accuracy of specific digits in your answer, but it will give you an idea of whether your exact answer is reasonable.

Answers to subtraction problems can be estimated by rounding each number to its largest place and then subtracting the rounded numbers. The estimate will be close to the actual answer.

EXAMPLE

Estimate the difference: 29,689 – 12,271. Then find the actual answer.

SOLUTION

Estimate	Actual
30,000	29,689
– 10,000	–12,271
20,000	17,418

▶ **Self-Check**

Estimate the difference: 39,872 – 6,819. Then find the actual answer.

Self-Check Answer

Estimate   33,000

Difference   33,053

Calculator Tip

Addition and subtraction can be done within the same problem. Subtract by pressing the minus sign.

Estimate the answer to each problem below and on the next page by rounding each number to its largest place and then subtracting. Then find the actual difference.

9.

Actual	Estimate
38,796	
– 25,432	

10.

Actual	Estimate
36,329	
–5,124	

11.

Actual	Estimate
26,392	
–9,684	

12.

Actual	Estimate
73,256	
–38,197	

Problems

13.

Actual	Estimate
41,630	
−7,814	

14.

Actual	Estimate
11,651	
−3,597	

15. What is the difference between a mortgage payment with a 7.80% rate and a payment with a 6.74% rate? Estimate the answer, then compute the exact answer.

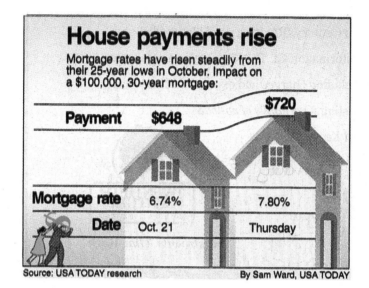

House payments rise

Mortgage rates have risen steadily from their 25-year lows in October. Impact on a $100,000, 30-year mortgage:

Payment	$648	$720
Mortgage rate	6.74%	7.80%
Date	Oct. 21	Thursday

Source: USA TODAY research

By Sam Ward, USA TODAY

Student Notes

SKILLBUILDER 1.11

Problem Solving: Business Applications

Learning Outcomes

After completing this skillbuilder, you should be able to:

- Solve business applications involving addition and subtraction.
- Solve a word problem using the problem-solving steps.

Using The Problem-Solving Steps

An important factor in solving word problems is knowing when to add, subtract, multiply, or divide. Often key words in the problem will indicate which operations are required. For example, *of, total,* and *how many in all* often indicate multiplication. *For each, per,* and *find the average* usually indicate division. *Plus, total, sum,* and *how many in all* often indicate addition. *Amount of increase, amount of decrease, less, difference, balance,* and *how many more* are words that often indicate subtraction. The word *what* indicates the unknown factor; *is* indicates equals. Examining these key words will help in expressing a word problem as a number problem.

To solve a word problem, read the problem carefully, more than once if necessary. Then follow these steps.

1. Determine what the problem is asking you to find.
2. Decide what information is necessary in order to solve the problem.
3. Decide what arithmetic operation to use (look for key words).
4. Write a problem and solve it.
5. **Reread** the problem to make sure that your answer is sensible. Then check the answer.

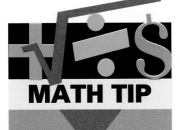

MATH TIP

When solving a word problem, be sure your answer is accurate. That is, check your work using the methods of checking you have learned. Then be sure your answer is sensible—for instance, a unit cost of $10 for a bicycle is not sensible and indicates an error in the solution process.

EXAMPLE

Elbert Products spent $1,800 on a holiday party for their employees, $2,100 on gifts for clients, $18,000 on employee bonuses, and $346 on holiday decorations. How much did they spend in all?

SOLUTION

1. What is the problem asking you to find?
   Total holiday expenses
2. What is the necessary information?
   Party: $1,800   Gifts: $2,100
   Bonuses: $18,000   Decorations: $346
3. Key Words: *in all*   (indicates addition)

4. Write a problem and solve it:

$$\begin{array}{r} \$1,800 \\ 2,100 \\ 18,000 \\ \underline{346} \\ \$22,246 \end{array}$$

5. Reread the problem and check by adding in reverse order.
　　Elbert Products spent $22,246 in all on holiday expenses.

 **Self-Check**

Five crates of parts are to be shipped by National Airways. The crates weigh 56 kg, 46 kg, 74 kg, 68 kg, and 60 kg. What is the total weight of the crates to be shipped?

Self-Check Answer

304 kg

Problems

Solve these problems using the problem-solving steps as a guide.

1. Daniel Durant, the maintenance plumber at the Sterling Apartments, had to replace damaged pipes in the sprinkler system. He cut and installed five pieces of pipe, measuring 8 ft, 3 ft, 2 ft, 9 ft, and 6 ft. How many feet of pipe in all are needed to repair the sprinkler system?

2. In 1992, 12,500,000 passengers used Main Terminal Air facilities. By 1995, the number had risen to 30 million passengers per year. What is the total increase in passenger use from 1992 to 1995?

3. A total of 5,828 people toured the plant facilities at South Service Company during their annual open house. This is 378 fewer people than toured during last year's open house. How many people toured last year?

Answers

1. _____

2. _____

3. _____

Problems

4. Fashion-tec has a production capacity of 35,000 units a week. Their actual production last week was Monday, 3,274 units; Tuesday, 2,492 units; Wednesday, 3,194 units; Thursday, 6,436 units; and Friday, 5,432 units. What was the difference between their actual production and their production capacity?

Answers

4. _____

Business Applications

Business transactions are often recorded in a record called an **account.** An account resembles the letter *T*. The left side of the account is called the **debit** side; the right side of the account is called the **credit** side. To find how much is in the cash account, add all the debit amounts; then add all the credit amounts. The smaller number is subtracted from the larger number. If the total of the debit side is larger, the account has a **debit,** or negative, **balance.** If the total on the credit side is larger, the account has a **credit,** or positive, **balance.**

EXAMPLE

Determine whether there is a debit or credit balance for debits of $160.00, $2,650.50, $65.45, $492.11, and $745.73 and credits of $56.25, $643.90, $85.00, and $234.75.

SOLUTION

Cash

Debit	Credit
$ 160.00	$ 56.25
2,650.50	643.90
65.45	85.00
492.11	234.75
745.73	$1,019.90
$4,113.79	

$4,113.79
−1,019.90
$3,093.89 Debit balance

Note that debits are greater than credits, so the balance is a debit balance.

 **Self-Check**

Determine whether there is a credit or debit balance for debits of $35.88, $1,405.90, $332.19, and $38.55 and credits of $1,089.34, $42.76, and $882.10.

Self-Check Answer

Cash

Debit	Credit
$ 35.88	$1,089.34
1,405.90	42.76
332.19	882.10
38.55	$2,014.20
$1,812.52	

$2,014.20 − $1,812.52 = $201.68 Credit balance

Problems

5. Add the debit and credit amounts in the accounts shown below. Subtract and determine whether the accounts have debit or credit balances.

Accounts Receivable		Supplies		Equipment		Accounts Payable	
Debit	Credit	Debit	Credit	Debit	Credit	Debit	Credit
5,678.90	790.00	150.00	50.00	7,000.00	1,000.00	1,025.00	5,200.00
790.00	2,678.90	65.00	6.75	950.00	450.00	340.00	680.00
1,434.08	434.08	93.56	12.80	10,450.25	4,050.00	340.00	575.50
53.25	500.00	29.04		700.00		2,600.00	590.00
2,000.00	53.25	6.85		3,450.50			3,090.75
194.34		42.09		1,268.60			1,258.05
53.25							

6. A cashier should make change in the fewest number of coins or bills possible. Use one dime, not two nickels. Use a $10 bill, not two $5 bills. Find the change for these purchases. Then select the correct coins and bills to make up the change.

	Amount Given In Payment	Amount of Purchase	Amount of Change
Ex.	*$10.00*	*$2.58*	*$7.42*
a.	5.00	1.49	_____
b.	7.00	6.43	_____
c.	20.00	5.04	_____
d.	15.00	12.51	_____
e.	1.00	.59	_____
f.	50.00	29.78	_____

1¢	5¢	10¢	25¢	$1	$5	$10	$20
2	*1*	*1*	*1*	*2*	*1*		

Business Applications

Calculator Tip

When solving word problems involving whole numbers, first estimate the answer. Then set the decimal selector to zero and work the problem on the calculator.

7. Hopkins Medical Equipment and Supplies showed the following sales and returns for two days of business. Monday: sales of $73.19, $25.40, $324.65, and $8.49 and returns of $20.77 and $56.82; Tuesday: sales of $38.89, $42.57, $17.33, and $168.55 and returns of $12.39. Determine the net sales for the two days. Did Hopkins Medical Equipment and Supplies have a debit or credit balance for the two days?

7. _____

8. Based on the art below, in which city would you want to own a home, and why?

8. _____

Where home prices rose and fell

Metro areas with the biggest changes in median prices of used homes the first quarter vs. 1993's first quarter:

Biggest increase
Biloxi-Gulfport, Miss. 19.1% to **$72,400**
Salt Lake City, Utah 19.0% to **$92,800**

Biggest decrease
Detroit   −8.4% to **$84,500**
Los Angeles −4.8% to **$188,400**

Source: National Association of Realtors

133-city list, 5B

By Cindy Hall and Bob Laird, USA TODAY

Student Notes

_____
_____
_____
_____
_____
_____
_____
_____

U N I T 2

WHOLE NUMBERS
Multiplication and Division

The U.S. government gathers data regularly about various aspects of our lives. Data are collected and analyzed for presentation through numerous publications, including almanacs. The partial chart to the right shows the amount of fuel consumed by different modes of ground transportation. In order to find averages, you need to divide.

Type of Vehicle	Fuel Consumption (billions of gallons)*	Avg. Fuel Used per Vehicle (gallons)*	Avg. Miles per Gallon*
Cars	72.4	505	21.00
Buses	0.9	1,436	5.36
Trucks	58.1	1,305	10.62

In this unit we study multiplication and division of whole numbers and money amounts. The business applications of multiplication and division that we consider include invoices, sales records, and averages.

Bank Manager

In the past decade or so, the banking industry has become quite competitive. In order to attract new customers to open an account, banks often run expensive radio, TV, and newspaper ads that may offer free gifts to new customers.

Jamie is a branch manager who is paid a base salary as well as a bonus, depending upon the profits of the branch. The bonus is an incentive to make the branch as profitable as possible. The manager must keep track of revenue and expenses in order to make appropriate decisions that will increase the profits (and paycheck). Jamie's duties often involve working with decimals.

Math Application

A particular bank is offering new customers a free toaster when opening an account. The manager predicts that 100 new accounts will be opened due to the offer. If each new account orders checks, from which the bank makes $8.00 profit, how much profit does the bank manager expect to make from the check orders?

$$100(\$8.00) =$$

The manager was right. The free gift offer resulted in 100 new accounts. If the free gift costs the bank $5.00 each, what is the profit the bank receives after the gifts have been purchased? Use the formula to determine the profit.

$$\$800 - 100(\$5) =$$

Critical Thinking Problem

The bank manager decides to give away coffee mugs, decorated with the bank's logo. Company A agrees to prepare the coffee mugs with the logo for $2.50 each. Company B agrees to prepare the coffee mugs for $1.50 each plus a one time set-up fee of $125.00. Which company do you think the manager should choose? (Hint: Think of the number of new accounts projected.)

SKILLBUILDER **2.1**

Multiplying Whole Numbers

Learning Outcomes

After completing this skillbuilder, you should be able to:

- Find the product of two whole numbers.
- Find the product of money and a whole number.

Multiplying Whole Numbers

The number by which you multiply is the **multiplier,** the number being multiplied is the **multiplicand,** and the result of the computation is the **product.**

When the multiplier has two or more digits, the multiplicand is multiplied separately by each digit in the multiplier. These products are then added to find the final product. It is important to align the subproducts properly so that the correct answer will be obtained when the products are added.

MATH TIP

When you find the product of 43 and 52, you first multiply 43 × 2 and then 43 × 50. Thus, the partial product in the second row is actually 2,150, although the zero is not usually written.

EXAMPLE

Find the product: 43×52.

SOLUTION

$$
\begin{array}{rl}
43 & \text{Multiplicand} \\
\times 52 & \text{Multiplier} \\
\hline
86 & \\
2\ 15 & \\
\hline
2{,}236 & \text{Product}
\end{array}
$$

 **Self-Check**

Find the product: 104×85.

Self-Check Answer

_____8,840_____

Problems

Find the products and any totals.

1. $41 \times 5 =$ _____	**2.** $359 \times 6 =$ _____	**3.** 78	***Answers***
$87 \times 4 =$ _____	$7{,}412 \times 9 =$ _____	$\times\,54$	1. _____
$44 \times 3 =$ _____	$432 \times 3 =$ _____		2. _____
$61 \times 7 =$ _____	$1{,}546 \times 7 =$ _____		3. _____
$58 \times 9 =$ _____	$435 \times 8 =$ _____		4. _____
$99 \times 2 =$ _____	$3{,}452 \times 4 =$ _____		5. _____
Total = _____	Total = _____		6. _____

4.   856
        $\times\,32$

5.   754
        $\times\,392$

6.   8,065
        $\times\quad44$

7.   6,543
        $\times\,798$

8.   842
       $\times\,53$

9. 421
      $\times\,365$

10. 5,384
       $\times\quad779$

11. 6,739
       $\times\quad504$

12. 9,126
       $\times\,867$

7. _____
8. _____
9. _____
10. _____
11. _____
12. _____

Find the products.

13.   1,288
        $\times\quad7$

14.   3,492
        $\times\quad5$

15.   658
        $\times\,36$

16.   16,928
        $\times\,1{,}007$

13. _____
14. _____
15. _____
16. _____

Multiplying With Money

When multiplying an amount of money given in dollars and cents by a whole number, the answer is also written in terms of dollars and cents.

Problems

Find the products.

Answers

17. $23.56	**18.** $54.75	**19.** $182.09	**20.** $265.28
× 8	× 12	× 28	× 309

17. _____

18. _____

19. _____

20. _____

21. Compute the extensions on the following invoice by multiplying the quantity times the unit price. What is the total amount due?

21. _____

**GUTFELD
HARDWARE
COMPANY**

234 Mandrake Avenue
Cleveland, OH 44118

Invoice Number 6308

To   Bochner Home Supply Company
        2654 Marston Road
        Lyndhurst, OH 44124

Date   May 24,19—
Terms 2/10, n/30
Via   Atlas Freight

	Quantity	Stock No.	Description	Unit Price	Amount
a.	24	ED–345	Electric drills—$3/8$ in.	$ 21.68	_____
b.	54	SS–46S	Socket wrench set—standard	6.50	_____
c.	18	SS–46M	Socket wrench set—metric	6.50	_____
d.	144	LP–10	Locking plier—10 in.	1.50	_____
e.	144	AW–8	Adjustable wrench—8 in.	2.28	_____
				Total	_____

Problems

Calculator Tip

To find the product of a number times itself on some calculators, enter the number and press $\times$ and $=$.

22. Krazee Karl's Record Shop purchased 2,116 record albums, tapes, and compact discs from a bankrupt record store for $2.25 each. Karl sold them in his own shop at $7.89 each for the 1,587 albums, and $8.57 each for the 312 tapes. The rest, which were compact discs, were sold for $12.39 each.
a. How much did Krazee Karl receive from the sale of the 2,116 pieces?

b. What was the gross profit realized on this sale (total sales – cost of goods = gross profit)?

23. Helena Betz used the Bosco Company car for a business trip of 456 miles. After the trip, the odometer on the company car read 30,524. What was the odomter reading before she left on the trip?

24. Look at the art below. If 25 people received the average refund, what would be their total refund amount for 1993?

Answers

22 a. _____

b. _____

23. _____

24. _____

Bigger tax refunds

Since 1989, the average income tax refund has increased 12.2%, vs. a 15.6% rise in the consumer price index.

Average refund

1989   **$916**

1993   **$1,025**[1]

1 - as of last Friday

Source: Internal Revenue Service   By Gary Visgaitis, USA TODAY

SKILLBUILDER 2.2
Checking and Estimating With Multiplication

Learning Outcomes

After completing this skillbuilder, you should be able to:

- Check the accuracy of multiplication by interchanging factors.
- Estimate products using rounded numbers.

Checking the Accuracy of Multiplication

To check the accuracy of multiplication, interchange the *factors* (the multiplier and the multiplicand) and multiply again. The product should be the same no matter what order the factors are multiplied.

MATH TIP

When multiplying two whole numbers, the product will be greater than the multiplier or multiplicand if neither is 1 or 0.

EXAMPLE

Multiply: 597×63. Check by interchanging factors.

SOLUTION

$$
\begin{array}{r}
597 \\
\times\ 63 \\
\hline
1\ 791 \\
35\ 82 \\
\hline
37{,}611
\end{array}
\qquad
\begin{array}{r}
\textit{Check} \\
63 \\
\times\ 597 \\
\hline
441 \\
567 \\
315 \\
\hline
37{,}611
\end{array}
$$

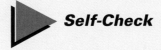

Self-Check

Multiply: 318×274. Check by interchanging the factors.

Self-Check Answer

___87,132___

Problems

Find the product for each of the following problems. Then check your answer by reversing the numbers and multiplying again. Show your work.

1. a.   *Check*   **b.**   *Check*   **c.**   *Check*   **d.**   *Check*

$\begin{array}{r} 95 \\ \times\,23 \\ \hline \end{array}$

$\begin{array}{r} 358 \\ \times\,74 \\ \hline \end{array}$

$\begin{array}{r} 716 \\ \times\,48 \\ \hline \end{array}$

$\begin{array}{r} 430 \\ \times\,852 \\ \hline \end{array}$

2. a.   *Check*   **b.**   *Check*   **c.**   *Check*   **d.**   *Check*

$\begin{array}{r} 3{,}725 \\ \times\,612 \\ \hline \end{array}$

$\begin{array}{r} 8{,}008 \\ \times\,907 \\ \hline \end{array}$

$\begin{array}{r} 4{,}500 \\ \times\,425 \\ \hline \end{array}$

$\begin{array}{r} 4{,}045 \\ \times\,222 \\ \hline \end{array}$

3. a.   *Check*   **b.**   *Check*   **c.**   *Check*   **d.**   *Check*

$\begin{array}{r} 521 \\ \times\,424 \\ \hline \end{array}$

$\begin{array}{r} 932 \\ \times\,283 \\ \hline \end{array}$

$\begin{array}{r} 654 \\ \times\,29 \\ \hline \end{array}$

$\begin{array}{r} 8{,}034 \\ \times\,419 \\ \hline \end{array}$

4. Here is a partial invoice. Check the invoice, and show the correct extensions and total.

Quantity	Unit Price	Extension
128 kg	$3.05/kg	$390.40
90 kg	1.19/kg	101.70
51 kg	0.51/kg	26.01
209 kg	0.81/kg	179.29
375 kg	1.10/kg	412.50
	Total	$1,109.90

Answers

_____

_____

_____

_____

_____

_____

_____

Problems

Estimating Answers in Multiplication

Calculator Tip

Estimation is a quick way of verifying the accuracy of a calculator computation.

Answers to multiplication problems can be estimated by rounding the multiplier and the multiplicand, usually to the first digit, and then multiplying the rounded numbers. The answer obtained will be close to the actual answer. Do not round single-digit numbers.

EXAMPLE

Multiply: 432 by 29.

SOLUTION

Estimate	*Actual*
400	432
× 30	× 29
12,000	12,528

 **Self-Check**

Estimate the product: 28,975 × 22. Then find the actual product.

Self-Check Answer

Estimate
600,000

Actual 637,450

Estimate the product. Then compute the actual answer.

5. a.

Actual	*Estimate*
178	
× 8	

b.

Actual	*Estimate*
295	
× 19	

c.

Actual	*Estimate*
1,003	
× 32	

d.

Actual	*Estimate*
650	
× 21	

e.

Actual	*Estimate*
994	
× 101	

Problems

6. Look at the chart below. If members in the IMSA pay $103 as dues, what is the total dues paid? Estimate the answer, then compute the exact answer.

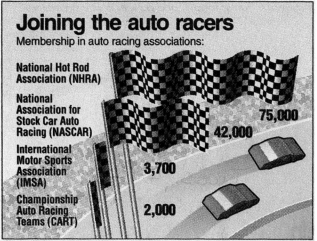

Joining the auto racers

Membership in auto racing associations:

National Hot Rod Association (NHRA)   75,000

National Association for Stock Car Auto Racing (NASCAR)   42,000

International Motor Sports Association (IMSA)   3,700

Championship Auto Racing Teams (CART)   2,000

Source: USA TODAY research   By Cindy Hall and Stephen Conley, USA TODAY

Student Notes

_____

_____

_____

_____

_____

_____

_____

_____

_____

_____

_____

_____

_____

_____

_____

_____

_____

_____

SKILLBUILDER 2.3

Business Applications for Multiplication

Learning Outcomes

After completing this skillbuilder, you should be able to:

■ Solve business applications involving multiplication.

Solving Business Applications

In many business applications involving multiplication, one of the factors is a money amount, written with a dollar sign and decimal point. When finding the product of an amount of money and a whole number, you must be sure to place the decimal point and dollar sign in the answer.

MATH TIP

You can find the product for $3.99 × 500 mentally by multiplying $4 × 500 and subtracting $0.01 × 500:
$4.00 × 500 = $2,000
$0.01 × 500 = $5.
$2000 − $5 = $1,995

EXAMPLE

Determine the wholesale cost of 485 sweaters if the wholesale unit price is $15.85.

SOLUTION

$$
\begin{array}{r}
\$ \quad 15.85 \\
485 \\
\hline
79\ 25 \\
1268\ 0 \\
6340 \\
\hline
\$7,687.25
\end{array}
$$
Place a dollar sign and decimal point in the answer

 **Self-Check**

Determine the wholesale cost of 500 picture frames if the wholesale unit price is $9.99.

Self-Check Answer

$4,995.00

Problems

Solve the following problems.

1. Midtown News Service inventory shows the following record of newspapers sold last week. What is the total of each newspaper sold and the total wholesale cost to Midtown?

Newspaper	Daily Sales							Total For Week	Wholesale Unit Price	Amount
	M	T	W	TH	F	SA	SU			
A	350	375	320	360	370	280	–0–	_____	$0.35	_____
B	500	511	485	470	462	370	–0–	_____	0.21	_____
C	185	197	221	184	170	152	–0–	_____	0.23	_____
D	212	249	248	240	225	185	–0–	_____	0.26	_____
E	408	417	441	416	409	397	–0–	_____	0.28	_____
F	160	173	181	173	168	160	–0–	_____	0.41	_____
G	–0–	–0–	–0–	–0–	–0–	–0–	402	_____	0.97	_____
H	–0–	–0–	–0–	–0–	–0–	–0–	508	_____	0.53	_____
									Total Cost	_____

Problems

Calculator Tip

If you use the calculator memory to store each subtotal in a calculation, you can find the total by using the memory recall feature.

2. Complete the following invoice.

Kaplan's Knits Inc.
124 Crescent Street
San Francisco, CA 94111

Invoice No. 385

To The Right Fit
3728 Harrison Street
Oakland, CA 94602

Date May 10, 19—
Terms 2/10. n/30
Via PIE

	Quantity	Item	Unit Price	Amount
a.	16	Women's knit suits	$ 160.00	_____
b.	18	Sweatshirts	52.00	_____
c.	24	Knit tops	12.00	_____
d.	40	Cotton knit shorts	9.75	_____
e.	30 doz	Knee socks/doz	21.50	_____
f.	180	T–shirts	8.00	_____
			Total	_____

3. Complete the following weekly sales report. Use any multiplication shortcuts that may be applicable.

			CLINT'S ELECTRONIC GAMES AND TOYS SALES WEEK OF FEBRUARY 23, 19—		
	Quantity	Unit	Item	Unit Selling Price	Total
a.	125	ea.	Super Video Game System	$198.97	_____
b.	25	ea.	Radio Control Racer	69.99	_____
c.	125	ea.	Game Software— Assorted	55.99	_____
d.	1,000	pkg	Batteries—C/D/9 V	2.49	_____
e.	50	ea.	Motorized Dinosaur	18.99	_____
f.	500	ea.	Magnetic Board Games	3.99	_____
				Total	_____

Problems

4. Complete the following invoice.

TALON'S CONSTRUCTION COMPANY
P.O. BOX 3729—BELVIDERE, NJ 07823

	Item No.	Quantity	Description	Price/Unit	Extension
a.	24657	1 ea.	Hi-Performance Vac.	$69.99	_____
b.	Stan10	7 ea.	Stain brush—1 in.	2.59	_____
c.	214SRSC	28 box	$6 \times 2\frac{1}{4}$ sheet rock screws	3.82	_____
d.	501-944	7 ea.	Exterior door	169.99	_____
e.	086-441	12 ea.	Window—3.2 × 4.5	152.60	_____
f.	6-002-182	25 shts	Plywood—⅝—CDX	16.79	_____
g.	6-308-324	24 bdl	Roof shingles—brown	12.50	_____
h.	6-004-048	3 rls	Roofing felt—15 lb	10.99	_____
i.	6-114-02	50 lb	Roofing nails	1.50	_____
				Total	_____

5. Look at the illustration below. If the increase between the 1996 and 2000 Summer Olympics is double that between 1992 and 1996, how many countries will participate in 2000?

Answer

5. _____

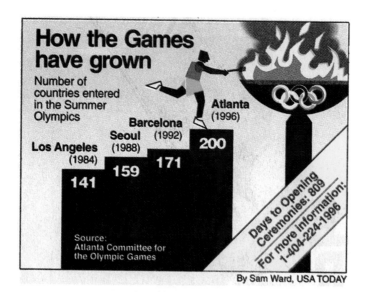

How the Games have grown

Number of countries entered in the Summer Olympics

Los Angeles (1984) 141
Seoul (1988) 159
Barcelona (1992) 171
Atlanta (1996) 200

Days to Opening Ceremonies: 809
For more information: 1-404-224-1996

Source: Atlanta Committee for the Olympic Games

By Sam Ward, USA TODAY

SKILLBUILDER 2.4

Dividing Whole Numbers

Learning Outcomes

After completing this skillbuilder, you should be able to:

■ Divide one whole number by another.

Dividing Whole Numbers

In division, the **dividend** is divided by the **divisor.** The result of the division process is the **quotient.** If the quotient is not a whole number, the fractional part can be expressed as a **remainder (R).** In the following example, the divisor (25) divides into the dividend (76) 3 times with a remainder of 1 (R1):

```
Quotient ─┐      ┌─Remainder
          ↓      ↓
             3 R1
Divisor → 25)76 ← Dividend
            − 75
              1
```

To divide, estimate how many times the first digit in the divisor goes into the first digit of the dividend. Write this estimate in the quotient and multiply it by the divisor. Then subtract the product from the dividend, as the example illustrates.

MATH TIP

To divide by 25 mentally, multiply the dividend by 4 and then divide the result by 100: 350 ÷ 25 = (350 × 4) ÷ 100 = 1400 ÷ 100 = 14.

EXAMPLE

Divide: 69 ÷ 34.

SOLUTION

```
    2 R1
34)69
   68
    1
```

6 ÷ 3 = 2; since we are dividing 69 by 34, write 2 above the 9.

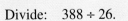

Self-Check

Divide: 388 ÷ 26.

Self-Check Answer

14 R 24

Problems

Find the quotient in each of the following problems. If the quotient is not a whole number, show what the remainder is.

Answers

1. 7)952 **2.** 4)1,684 **3.** 5)2,075

1. _____

2. _____

3. _____

4. 3)2,474 **5.** 16)592 **6.** 43)9,419

4. _____

5. _____

6. _____

7. 633)187,564 **8.** 458)219,158

7. _____

8. _____

9. 17)34,051 **10.** 18)63,018

9. _____

10. _____

Problems

Answers

Calculator Tip

When you do a division problem on your calculator, the result will be given as a decimal. For example, enter 69 ÷ 34 =.
The result will be 2.0294117. To find the remainder, subtract 2 and multiply the result by 34. (As a result of rounding, the remainder will not be exact.)

11. $34\overline{)64,038}$

12. $123\overline{)426,441}$

11. _____

12. _____

13. $45,968 \div 26$

14. $1,562 \div 502$

13. _____

14. _____

15. $344,527 \div 13$

16. $17,529 \div 83$

15. _____

16. _____

17. $547 \div 312$

18. $65,952 \div 109$

17. _____

18. _____

19. Look at the graph below. If the tax increase was cut in half, how much more would a couple earning $300,000 pay in 1993 than in 1992?

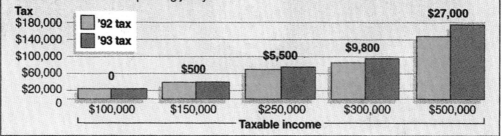

Wealthy feeling tax bite

Wealthy people are seeing their tax bills go up by thousands of dollars as President Clinton's new tax law kicks in. The top rates jumped to 36% and 39.6% from 31%. Here's what that means for a married couple filing jointly.

Tax
- '92 tax
- '93 tax

$180,000
$140,000
$100,000
$60,000
$20,000
0

$27,000

$9,800

$5,500

$500

0

$100,000 $150,000 $250,000 $300,000 $500,000

Taxable income

Source: Ernst & Young, USA TODAY research

By Elys A. McLean, USA TODAY

Student Notes

_____

_____

_____

_____

_____

_____

_____

_____

_____

_____

_____

_____

_____

_____

_____

_____

_____

_____

_____

_____

_____

_____

SKILLBUILDER 2.5

Checking and Estimating With Division

Learning Outcomes

After completing this skillbuilder, you should be able to:

- Check the accuracy of division problems using multiplication.
- Estimate the answer to division using rounded numbers.

Checking Division

To check the accuracy of a division computation, multiply the divisor by the quotient and add any remainder. The result should equal the dividend.

$$(\text{Divisor} \times \text{Quotient}) + \text{Remainder} = \text{Dividend}$$

MATH TIP

When dividing whole numbers, the quotient will be less than the dividend unless the divisor is 1.

EXAMPLE

Divide: 2,586 ÷ 26. Check your answer.

SOLUTION

$$
\begin{array}{r}
99 \text{ R12} \\
26)\overline{2,586} \\
2\ 34 \\
\hline
246 \\
234 \\
\hline
12
\end{array}
\qquad
\begin{array}{r}
\textit{Check} \\
99 \\
\times 26 \\
\hline
594 \\
1\ 98 \\
\hline
2\ 574 \\
+12 \\
\hline
2,586
\end{array}
$$

▶ **Self-Check**

Divide: 3,427 ÷ 23. Check your answer.

Self-Check Answer

____149____

Problems

Find the quotient in each problem, and indicate what the remainder is, if any. Check your answers by the method shown in the preceding example. Show the details of your work.

	Check		*Check*		*Check*	*Answers*
1. 62)3,658		**2.** 38)5,780		**3.** 84)9,660		1. _____
						2. _____
						3. _____

	Check		*Check*		*Check*	
4. 75)22,726		**5.** 5)251,530		**6.** 42)11,258		4. _____
						5. _____
						6. _____

	Check		*Check*		*Check*	
						7. _____
7. 48)3,985		**8.** 77)15,864		**9.** 36)1,730		8. _____
						9. _____

Estimating Answers in Division

Answers to division problems can be estimated by rounding the divisor and the dividend, usually to the first digit. It is not necessary to round the divisor if it is a single-digit number, .

When the second digit of the number is 5 or more, it is usually more accurate to round off to the second digit instead of the first. For example, you can more accurately estimate 17,784 ÷ 152 by finding 18,000 ÷ 150. It is also helpful to use *compatible* numbers—numbers that can be divided easily. For example, to estimate 14,124 ÷ 428, use 12,000 ÷ 400, because 12 and 4 are compatible.

Problems

EXAMPLE

Estimate: 6,192 ÷ 48. Find the actual answer.

SOLUTION

Actual	Estimate
129	120
48)6,192	50)6,000
4 8	
1 39	
96	
432	
432	

Self-Check

Estimate: 1,240 ÷ 8. Then find the actual answer.

Self-Check Answer

Estimate ___150___

Actual ___155___

Estimate the answers. Then find the actual answers. Show the details of your work.

10. *Actual* *Estimate*

42)8,294

11. *Actual* *Estimate*

28)92,540

12. *Actual* *Estimate*

428)14,124

Calculator Tip

Estimate the answer to a division problem before solving it with a calculator.

13. *Actual* *Estimate*

91)2,184

14. *Actual* *Estimate*

68)3,048

15. *Actual* *Estimate*

29)9,396

Problems

16. Look at the chart below. How many pages of the GATT does it take to weigh a pound?

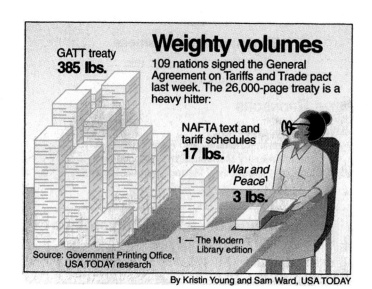

Weighty volumes

109 nations signed the General Agreement on Tariffs and Trade pact last week. The 26,000-page treaty is a heavy hitter:

GATT treaty **385 lbs.**

NAFTA text and tariff schedules **17 lbs.**

War and Peace[1] **3 lbs.**

1 — The Modern Library edition

Source: Government Printing Office, USA TODAY research

By Kristin Young and Sam Ward, USA TODAY

Student Notes

_____

_____

_____

_____

_____

_____

_____

_____

_____

_____

_____

_____

_____

_____

_____

_____

_____

_____

_____

SKILLBUILDER 2.6

Business Applications for Division

Learning Outcomes

After completing this skillbuilder, you should be able to:

■ Solve business applications involving division.

Solving Business Applications

Many products are priced and sold in units of 100 or 1,000. The abbreviation for 100 units is C, and the abbreviation for 1,000 units is M. A hundredweight, or 100 lb, is abbreviated cwt. The abbreviation for 1,000 board feet of lumber is MB or MBF. Other abbreviations you may encounter are doz (dozen) and gr (gross, or 144 pieces).

MATH TIP

To divide by powers of 10 (10, 100, 1,000, etc.), drop the same number of digits, starting at the right, as you have zeros in the power of 10. The dropped digits will be the remainder. For example: 9,427 ÷ 10 = 942 R7 and 35,600 ÷ 100 = 356 (R0).

EXAMPLE

An office supply store sells envelopes at $2.50/C. Find the cost of 600 of these envelopes.

SOLUTION

First we need to determine the number of 100 units in 600 envelopes:

$$\frac{600}{100} = 6$$

600 envelopes at $2.50/C → 6 × $2.50 = $15.00
The envelopes cost $15.00

Self-Check

Talon's Lumber Yard bought 8,000 board feet of lumber for $27/MBF. What was the cost of the lumber?

Self-Check Answer

$216

Problems

Solve the following problems.

Answers

1. A grocery supplier sold 900 lb of sugar for $33.50/cwt. What did the supplier charge for the sugar?

1. _____

2. Clint's Lawn Service bought 8,000 kg of potash for $159/M kg. What was the charge listed on the invoice?

2. _____

3. A nursery ordered 720 packets of flower seeds priced $57.60/gr. What was the cost of the flower-seed packets?

3. _____

Complete the extensions on each of the following invoices. Round to the nearest cent, where necessary.

4.

Gladeau Garden Supply
212 State Highway 30
Hohokus, New Jersey 07423

Invoice No. 11857

To   The Green Thumb
1123 East End Road
Weehawken, NJ 07087

Date   March 3, 19—
Terms C.O.D.
Via   Pickup

	Quantity	Item	Unit Price	Amount
a.	8,000 kg	# 18–216 Weed block	$ 909/M kg	_____
b.	1,200 lb	# 18–128 Green seed	133.33/cwt	_____
c.	800 bags	# 18–861 Peat grow—40 lb	80.00/C bags	_____
d.	3,000 yd	# 18–887 Burlap	970/M yd	_____
			Total	_____

Problems

5.

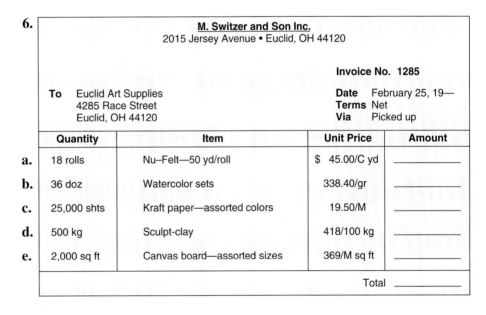

Milton Paper Supply
843 Ridgedale Avenue
Cedar Knolls, PA 18350

Invoice No. 12289

To   Rainbow Caterers
4298 Mt Pleasant Avenue
Livingwell, NJ 07793

Date   May 1, 19—
Terms Net
Via   Delivered

	Item	Quantity	Description	Unit Price	Extension
a.	11–852	10 M	Napkins—economy	$ 3.71/M	_____
b.	41–725	250 C	Plates—5 in.	2.87/C	_____
c.	41–730	110 C	Plates—10 in.	5.21/C	_____
d.	58–0710	480 doz	Cutlery sets	21.60/gr	_____
e.	29–365	5 L	Liquid Kleen concentrate	2.89/L	_____
				Total	_____

6.

M. Switzer and Son Inc.
2015 Jersey Avenue • Euclid, OH 44120

Invoice No. 1285

To   Euclid Art Supplies
4285 Race Street
Euclid, OH 44120

Date   February 25, 19—
Terms Net
Via   Picked up

	Quantity	Item	Unit Price	Amount
a.	18 rolls	Nu–Felt—50 yd/roll	$ 45.00/C yd	_____
b.	36 doz	Watercolor sets	338.40/gr	_____
c.	25,000 shts	Kraft paper—assorted colors	19.50/M	_____
d.	500 kg	Sculpt-clay	418/100 kg	_____
e.	2,000 sq ft	Canvas board—assorted sizes	369/M sq ft	_____
			Total	_____

7.

M. Switzer and Son Inc.
2015 Jersey Avenue • Euclid, OH 44120

Invoice No. 1292

To   Hopewell Enterprises
853 Spinning Road
Dayton, Ohio 45431

Date   April 12, 19—
Terms C.O.D.
Via   Lee's Express

	Quantity	Item	Unit Price	Amount
a.	48 cases	Duplicating paper—5M shts/cse	$ 9.78/M shts	_____
b.	6 gr	Ballpoint pens	1.29/doz	_____
c.	4 cases	Toner—10 kg/cases	15.98/kg	_____
d.	12 doz	Mylar ribbon cartridges	228.96/gr	_____
e.	11.5 M	Envelopes—#10	4.85/C	_____
			Total	_____

Problems

8.

		Seoul Electronics International			
		Mount Olive Foreign Trade Zone			

Invoice No. 11840

To Sight and Sound
1128 Bay Street
Seattle, WA 08593

Date April 17, 19—
Terms 5/10, 2/30, n/45

	Item	Quantity	Description	Unit Price	Amount
a.	VR2400/6267	38	VCR—remote/digital	$ 219.43	_____
b.	DX610/6312	60	Compact disc player	56.99	_____
c.	8200/3078	75	Digital car stereo	56.98	_____
d.	81/TRI6	150	Speaker—Triaxial	17.01	_____
e.	1955/3108	20	Color TV—remote 19 in.	170.43	_____
f.	SCA–630	4	200-watt stereo system	455.43	_____
					_____
				Subtotal	_____
				Freight	_____
				Total	_____

9. Which 2 quarters seem to be the best, each year, based on 1992 and 1993?

Answer

9. _____

Calculator Tip

To solve a problem such as Problem 3, enter 720 ÷ 144 × 57.60 = .

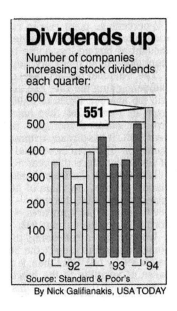

Dividends up

Number of companies increasing stock dividends each quarter:

551

Source: Standard & Poor's
By Nick Galifianakis, USA TODAY

SKILLBUILDER 2.7

Computing Averages

Learning Outcomes

After completing this skillbuilder, you should be able to:

■ Compute the average of a group of numbers.

Computing Averages

The **average** of a group of numbers is found by dividing the sum of the set of numbers by the number of items in the set. The average is also called the **mean** of the numbers.

MATH TIP

When computing averages, the solution is reasonable if it is between the lowest and highest amounts being added.

EXAMPLE

During a 3-year period, a sales representative earned $28,900; $32,450; and $38,526. Find the average annual income for this period.

SOLUTION

$28,900
  32,450
  38,526
————
$99,876   Total 3-year income

$99,876 ÷ 3 = $33,292   Average annual income

 **Self-Check**

Initial investments in businesses in Coulterville were $36,645; $51,000; $44,500; $50,125; and $55,300. Find the average investment.

Self-Check Answer

$47,514

Problems

Solve the following problems.

1. During this year's United Campaign, 112 communities in Morris County contributed a total of $4,974,144. What was the average contribution per community?

1. _____

2. Four sales representatives employed by Randolph Realty Associates earned $4,220; $2,965; $3,763; and $5,175 during the past month. What was the average of their monthly earnings?

2. _____

3. Marge Pineros is a sales representative with Randolph Realty Associates. For the past 12 months, her earnings were $3,364; $3,695; $3,215; $4,150; $2,915; $4,197; $3,496; $3,751; $3,215; $3,935; $3,202; and $3,525.
a. What were her total earnings for the year?
b. What were her average monthly earnings to the nearest cent?

3. **a.** _____

   b. _____

4. The nine sales representatives for Bledsoe Products had expense accounts for the past 3 months as follows: $1,128; $1,374; $801; $1,506; $1,251; $888; $1,029; $1,317; and $1,533.
a. What was the average expense account for the 3-month period?
b. What was the average monthly expense account?

4. **a.** _____

   b. _____

Problems

Calculator Tip

Use the average key or the counting key to compute simple averages. Press the average key after the sum of a series of numbers is found. Press the counting key and then divide by the number given.

5. Stella Chandler uses her car for business. For the past year, she drove the following mileage. Write the total cost of gasoline per quarter and the year in the Total Cost column.

Quarter	Miles	Gallons of Gas Used	Price Per Gallon	Total Cost
Jan.–Mar.	1,806	111	$1.23	_____
Apr.–June	3,015	196	$1.22	_____
July–Sept.	3,674	240	$1.20	_____
Oct.–Dec.	2,870	163	$1.21	_____
Total	_____	_____		_____

a. What was her average mileage for a gallon of gas? Round your answer to the nearest hundredth.

b. What was the average price she paid for a gallon? Round off your answer to the nearest cent.

6. A jeweler sold the following numbers of rings each week during a recent month: 93, 84, 90, 77. What was the average number of rings sold each week?

7. The weekly commissions for Greg Restik, a furniture salesperson, for the past 4 weeks were $1,000, $850, $1,200, and $900. What was his average commission per week?

8. Naomi Melesco, a produce actioneer, took a total of 3,000 bags of potatoes to auction. Of the 3,000 bags, 40 bags were spoiled and were dumped prior to the sale. 960 bags were sold for $6.75 per bag; 1,100 bags for $6.25 per bag; 500 bags for $7.15 per bag; and the remainder for $7.10 per bag. What was the average price per bag?

9. Long's Lilies bought 3,500 planters for $3,500 at a bankruptcy auction. It sold 1,850 planters for $2.70 each, and 1,000 for $2.85 each. For how much would Long's Lilies have to sell the balance (to the nearest dollar) in order to realize the average selling price of $3 a planter?

Answers

5. a. _____

   b. _____

6. _____

7. _____

8. _____

9. _____

U N I T 3

DECIMALS

Addition, Subtraction, Multiplication, and Division

The chart shown here gives total expenditures for advertisers by category and medium. The expenditures are given in decimals, in millions of dollars. Why do you think this is the case? Which medium attracted the most advertising? In which category was the most money spent? What factors might influence the media in which a company would choose to advertise?

In this unit we study operations with decimal numbers, as well as rounding decimals and shortcuts in multiplying and dividing decimals.

TOTAL U.S. AD SPENDING BY CATEGORY AND MEDIUM, 1991

Source: Advertising Age, Sept. 23, 1992 © Crain Communications Inc. 1992								
Category	Total Ad spending	Magazines	Sunday magazines	Newspaper	Network TV	Spot TV	Syndicated TV	Cable TV	Network radio
Automotive	$5,259.1	$940.3	$38.3	$746.2	$1,559.5	$1,480.6	$120.7	$117.7	$72.2
Retail	5,125.9	197.1	93.0	2,491.7	351.7	1,528.0	40.1	42.9	92.1
Business, consumer serv.	3,582.8	444.3	28.4	1,207.2	598.3	811.2	72.8	112.2	77.2
Food	3,551.9	433.4	29.1	36.5	1,443.8	974.0	346.1	141.0	41.9
Entertainment	2,913.2	58.5	36.7	387.3	802.3	1,196.0	196.1	83.0	15.6
Toiletries & cosmetics	2,249.3	631.8	21.9	6.8	958.6	293.3	215.3	92.5	8.2
Travel & hotels	2,123.2	335.4	49.1	1,064.5	212.7	233.6	8.6	49.1	33.4
Drugs & remedies	1,807.9	164.8	19.5	78.2	851.6	341.4	171.3	80.2	56.3
Direct response cos.	1,192.9	555.7	302.3	65.1	39.7	76.6	33.6	53.2	62.1
Candy, snacks & soft drinks	1,146.2	55.2	2.4	13.5	434.3	318.3	185.1	72.2	20.5
Apparel, footwear	904.8	416.0	21.5	7.0	235.4	88.5	58.9	41.0	7.5
Beer & wine	839.6	51.6	5.4	12.8	347.0	209.2	60.1	43.5	1.5
Insurance & real estate	705.3	141.6	8.9	187.3	171.5	106.2	9.1	19.0	19.1
Publishing & media	700.4	195.7	11.8	194.1	40.3	137.7	12.1	19.1	21.9

Supermarket Manager

A supermarket manager supervises the employees, plans schedules, places food orders, and sets prices. Each of these functions is important to the store's success. The decisions made in hiring, firing, and scheduling are basically a matter of the manager using good judgment based on experience. However, to order food to stock shelves and to set prices for all the items, a manager must use math skills. Knowing the competition and understanding manufacturers' requirements are essential.

Math Application

A store manager wants to estimate the weekly cost of meat for a family of 2 adults and 2 children for next week's ad. He chooses 6 pounds of ground beef ($2.09/lb), a 5-pound pork roast ($2.89/lb), one package of hot dogs ($1.49/pkg), and 3 cans of tuna ($.85/can). What is the manager's estimate?

 ## Critical Thinking Problem

The store manager decides to sell a gallon of milk for $.80 below cost in order to lure customers into the store. What are some ways that the manager might try to regain the loss on the milk?

NAME: _____

DATE: _____

SKILLBUILDER 3.1

Reading and Writing Decimals; Rounding Decimals

Learning Outcomes

After completing this skillbuilder, you should be able to:

- Read and write decimals using words.
- Read word names of decimals and write decimal names using digits.
- Round decimals to a specified place.

Reading and Writing Decimals

A **decimal number,** or **decimal,** is a number containing a decimal point and digits to the right of the decimal. The digits to the right of the decimal point are called **decimal fractions.** Decimal fractions have a value of less than one. Each place in a decimal fraction is one tenth the value of the place to its left.

A digit in the hundredths place has a value of one-tenth of the same digit in the tenths place. A digit in the thousandths place has a value of one-tenth of the same digit in the hundredths place.

To read or write a decimal fraction in words, read the digits as if they were a whole number and then read the name of the place of the digit on the extreme right. If the decimal has a whole number part, read the number to the left of the decimal point. Next, read the decimal point as *and.* Finally, read the name of the decimal fraction.

Look at the Example and Solution shown on the next page.

Problems

MATH TIP

Remember to place commas properly when writing decimals. Do not use the word *and* when reading the whole number part—*and* is read only when a decimal point occurs.

EXAMPLE

Read 203.167 and then write it in words.

SOLUTION

Write the number in a place-value chart.

So, 203.167 is read "two hundred three and one hundred sixty-seven thousandths."

 **Self-Check**

Read the number 0.5892 and write it in words.

Self-Check Answer

Five thousand eight hundred ninety-two ten thousandths

Read these numbers and then write them in words.

Answers

1.	0.7	_____
2.	0.23	_____
3.	0.68	_____
4.	4.016	_____
5.	5.118	_____
6	17.83	_____
7	24.216	_____
8.	3.0808	_____
9.	25.2006	_____
10.	4.00006	_____
11.	379.058	_____
12.	8,016.09	_____

Problems

Writing Decimals Using Digits

A place-value chart can also be used to help write the word name of a decimal using digits.

EXAMPLE

Write the number four hundred five ten thousandths using digits.

SOLUTION

Write the number 405 in a place-value chart, placing the 5 in the ten thousandths place. Zeros must be added to any empty places between the number and the decimal point.

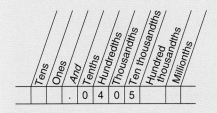

Four hundred five ten thousandths is written 0.0405. Notice that we write a 0 in the units place when writing decimal fractions.

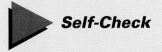 **Self-Check**

Read the number sixteen hundredths and write it using digits.

Self-Check Answer

_____0.16_____

Read these numbers, and then write them using digits.

Answers

13. Three hundred seventy-three thousandths — 13. _____

14. Eighty-four ten thousandths — 14. _____

15. Forty-seven hundredths — 15. _____

16. Seven tenths — 16. _____

17. Thirty-one and six thousandths — 17. _____

18. One thousand, sixty and six tenths — 18. _____

19. Twenty-seven and thirteen hundredths — 19. _____

20. Three and nine hundred nine ten thousandths — 20. _____

21. One and sixteen thousandths — 21. _____

22. One and nine hundred thousandths — 22. _____

Problems

Write the following numbers using digits in the columns at the right. Align them correctly both horizontally and vertically.

Answers

23. **a.** One hundred sixteen

    b. Five and six thousandths

    c. Thirty-nine

    d. Fifty-five and one hundred fifty-one thousandths

    e. Eighty-nine and seven tenths

24. **a.** Two thousand, sixty three and four tenths

    b. Sixty-six and sixty-six hundredths

    c. Thirty-five and five tenths

    d. Six thousand, three hundred twenty-seven

    e. Five hundred fourteen

25. **a.** Twenty-seven and three hundred fourteen ten thousandths

    b. Seventeen and seventeen thousandths

    c. Forty-seven and forty-seven hundredths

    d. Nine thousand seven hundred sixty-two ten thousandths

    e. Twenty-nine and eight tenths

26. **a.** Nine hundred seventy-two and one hundred twenty-five thousandths

    b. Forty-three and forty-three thousandths

    c. Fifty-two and five tenths

    d. Three and three thousandths

    e. Seven hundred six and sixty-seven thousandths

23. **a.**
    b.
    c.
    d.
    e.

24. **a.**
    b.
    c.
    d.
    e.

25. **a.**
    b.
    c.
    d.
    e.

26. **a.**
    b.
    c.
    d.
    e.

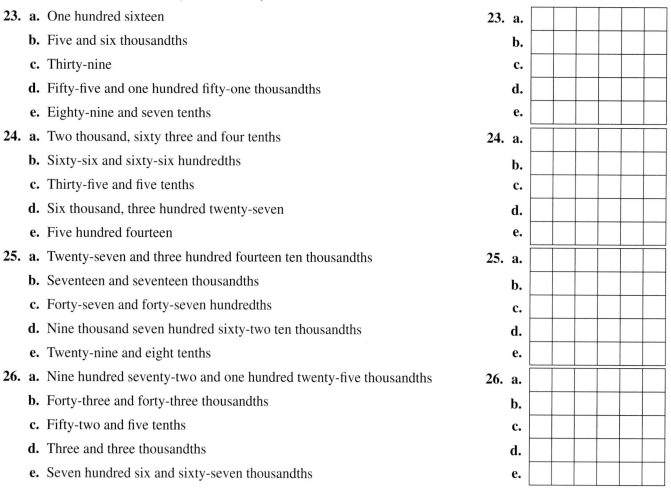

Rounding Decimals

Rounding decimals to a specified place is very similar to rounding whole numbers. To round a number to a specified place, identify the digit in the rounding place. Then look at the digit to the right of that place. If it is 5 or more, increase the digit in the rounding place by 1 and drop the digits to the right of the rounding place. If the digit is less than 5, drop the digits to the right of the rounding place.

Problems

EXAMPLE

a. Round 2.7063 to the nearest tenth.
b. Round $104.725 to the nearest cent.

SOLUTION

a. Place a line under the digit in the tenths place. Then place an arrow over the digit to the right.

$$\downarrow$$
2.7063

Since 0 is less than 5, drop the digits to the right of 7.

2.7063 to the nearest tenth is 2.7.

b. Place a line under the digit in the cents place. Then place an arrow over the digit to the right.

$$\downarrow$$
$104.725

Since 5 is 5 or greater, increase the cents digit by 1 and drop the 5.

$104.725 to the nearest cent is $104.73.

▶ **Self-Check**

Round $39.6027 to the nearest cent.

Self-Check Answer

$39.60

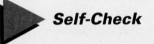

Calculator Tip

Most calculators with rounding selectors are set on the 5/4 position except when finding unit price (any fraction of a cent is passed on to the consumer).

Round to the specified place.

27. 53.79, nearest tenth

28. 0.9432, nearest thousandth

29. $16.827, nearest cent

30. $12.695, nearest cent

31. 8.995, nearest hundredth

32. $0.864, nearest cent

33. 2,756.8051, nearest hundredth

34. $1.555, nearest cent

Answers

27. ———————

28. ———————

29. ———————

30. ———————

31. ———————

32. ———————

33. ———————

34. ———————

Business Applications

35. Kelso Enterprises sells merchandise by catalog. They include a chart that shows the shipping charges, which are determined by the dollar amount of the order. Use the table to determine the shipping charge for each order.

SHIPPING AND HANDLING	
Order Amount	**Shipping Charge**
Up to $15.00	$3.75
$15.01 –$30.00	$5.35
$30.01 –$50.00	$6.35
$50.01 –$80.00	$6.95
$80.01–$100.00	$7.55
$100.01–$200.00	$7.95
Orders over $200.00	$8.25

a. $ 8.25
b. $ 45.00
c. $ 73.40
d. $198.00
e. $ 26.30
f. $505.34
g. $ 97.50
h. $ 18.00
i. $201.00
j. $ 30.02

Answers

35. a._____
   b._____
   c._____
   d._____
   e._____
   f._____
   g._____
   h._____
   i._____
   j._____

36. Shelves and More computed the average cost of a shelf unit to be $284.7961. What is the average cost to the nearest cent?

36. _____

NAME: _____

DATE: _____

SKILLBUILDER 3.2
Adding Decimals

After completing this skillbuilder, you should be able to:

■ Add two or more decimals.

Adding Decimals

When adding decimals, place the addends in a column with the decimal points aligned. Recall that annexing zeros to the right of a decimal fraction or to the right of the decimal point in a whole number does not change the value of the decimal. It may be helpful to annex zeros to help align columns properly. Remember to place the decimal point in the sum directly below the decimal points in the addends.

MATH TIP

The decimal point is understood in a whole number; therefore, the decimal point does not have to be written. Remember, however, to put the decimal point to the right of the number when solving problems.

EXAMPLE

find the sum:
425.04 + 22 + 9.132 + 0.37.

SOLUTION

$$
\begin{array}{r}
\overset{1}{~}\quad\overset{1}{~}\quad\quad \\
425.040 \leftarrow \\
22.000 \leftarrow \\
9.132 \\
0.370 \leftarrow \\
\hline
456.542
\end{array}
$$
← Annex zeros

 **Self-Check**

find the sum: 0.34 + 23.901 + 634 + 0.8

Self-Check Answer

_____659.041_____

Problems

Find each sum.

1.	621.88 + 413.93	**2.**	28.51 + 3.18	**3.**	46.94 + 31.27

4.	5.79 + 266.66	**5.**	$14.35 + 25.99	**6.**	$545.10 + 22.60

7.	7,190.9 15.56 0.042 + 123.25	**8.**	61.188 115.75 0.0125 + 9.9	**9.**	17.471 36.31 257.225 + 15.9

10.	$5,127.75 73 125.25 + 2,800.00	**11.**	$317 12.26 416.97 + 9.33	**12.**	$ 948.31 130.40 3,289.14 + 51.76

Answers

13. 37.83 + 115.7 + 25.678   **13.** _____

14. 52.15 + 74.716 + 1.12   **14.** _____

15. 14.7 + 6.017 + 83.49 + 9   **15.** _____

16. $75.34 + $13.50 + $215.49 + $5.00   **16.** _____

Problems

17. $14.5 + 56.78 + 19.316 + 10$

17. _____

18. $4.971 + 24.72 + 220 + 1.003$

18. _____

19. $13 + 127.7 + 263.4176 + 17.22$

19. _____

20. $\$415.27 + \$51.49 + \$12.45 + \$1,243.00$

20. _____

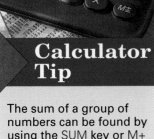

Calculator Tip

The sum of a group of numbers can be found by using the SUM key or M+ key. Press [SUM] or [M+] after entering each number. To access the total, press [RCL] or [MR].

Business Applications

21. Complete the following invoice.

Date:	5/17			Total Amount:	
Sales No.:	005444				
Salesperson:	J. Smith				

Quantity	Stock No.	Description	Unit	Amount
3 boxes	372-A	Syringes (1 doz.)	$10.25	_____
4 boxes	564-C	Cotton balls	$10.68	_____
2 boxes	119-B	$2\frac{1}{2}$-in. elastic bandages	$12.34	_____
			Total	_____

22. St. Mary's Hospital and Medical Center sent out the following bill. Compute the total due.

Patient: Doris Kwan

Service Description	Charges
Pharmacy	$ 50.90
Laboratory pathology	41.00
Operating room	658.00
Anesthesiology	321.00
Respiratory therapy	17.00
Recovery room	155.00
Electrocardiogram	56.00
Total:	_____

Business Applications

23. Aaron Goldman recorded his expenses on a business trip for Alco Electronics. Complete his travel expense report.

Expense Item	Su 4/27	M 4/28	T 4/29	W 4/30	Th 5/1	F 5/2	Sa 5/3	Totals
Meals			$32.00	$32.00	$32.00	$32.00		_____
Mileage			35.00					_____
Lodging			97.34	97.34	97.34	97.34		_____
Tips			6.00	7.00	6.00	7.00		_____
Taxi			17.00			17.00		_____
Tolls			1.69			1.69		_____
Airfare			212.00					_____
Parking						10.00		_____
Daily totals			_____	_____	_____	_____		_____

24. If you bought 1 gallon each of liquid propane and compressed natural gas, what would be the total cost?

24. _____

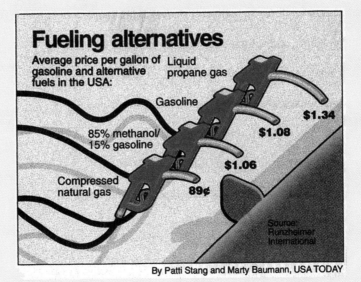

Fueling alternatives

Average price per gallon of gasoline and alternative fuels in the USA:

Liquid propane gas — $1.34

Gasoline — $1.08

85% methanol/15% gasoline — $1.06

Compressed natural gas — 89¢

Source: Runzheimer International

By Patti Stang and Marty Baumann, USA TODAY

Student Notes

SKILLBUILDER 3.3
Subtracting Decimals

Learning Outcomes

After completing this skillbuilder, you should be able to:

■ Find the difference of two decimals.

Subtracting Decimals

To find the difference between two decimals, write the smaller decimal below the greater one, lining up the decimal points. Annex zeros as needed so that each decimal has the same number of places. Remember to write the decimal point in the difference directly below the decimal points in the problem.

MATH TIP

Parentheses are generally used to indicate negative numbers in business. Thus, a loss of $2,000 is written ($2,000).

EXAMPLE

Find the difference: 156 − 1.799.

SOLUTION

```
       9 9
   5 10 10 10
  156.000   ← Annex zeros so that 156 has three decimal places
    1.799
  154.201
```

 **Self-Check**

Find the difference: 38.47 − 0.233.

Self-Check Answer

___38.237___

Problems

Find the difference.

1.   593.
$\underline{-78.63}$

2.   437.712
$\underline{-221.353}$

3.   848.40
$\underline{-78.216}$

4. $88.60
$\underline{-71.70}$

5. $124.95
$\underline{-86.78}$

6. $349.57
$\underline{-128.35}$

7.   856.7
$\underline{-46.523}$

8.   76.
$\underline{-0.437}$

9.   401.
$\underline{-9.631}$

10. $936.00
$\underline{-341.49}$

11. $2,679.00
$\underline{-1,538.17}$

12. $3,213.56
$\underline{-1,479.57}$

13.   141.21 − 69.90

14. 246.38 − 78.83

15.   161.2 − 66.3

16. 721.9 − 88.33

17.   441.4 − 215.9

18. 44.27 − 0.6124

19. $621.66 − $78.43

20. $7,000 − $4,833.63

Answers

13. _____

14. _____

15. _____

16. _____

17. _____

18. _____

19. _____

20. _____

Calculator Tip

Negative balances on a calculator are shown by the minus sign at the left.

Business Applications

21. Complete the following deposit slip.

Currency	$352.00
Coins	3.75
Checks (list separately)	33.40
	129.25
	18.99
Subtotal	_____
Less cash received	$ 50.00
Total deposit	_____

22. St. Mary's Hospital and Medical Center sent a statement to Brad Hutchens following his surgery. Complete the computations to give the current balance. The patient portion is any amount remaining after insurance payments and adjustments.

St. Mary's
Hospital and Medical Center
Patient: Brad Hutchens

Account #: **4302719**
Statement Date: **5/17/—**

Date	Description of Service	Amount
7/17/—	Beginning balance	$1,298.90
8/07/—	Insurance adjustment	194.98
8/07/—	Payment/Insurance company	883.94
	Current balance	_____
	Patient portion	_____
	Total account balance	_____

Business Applications

23. Hal's Fish and Tackle prepared a report showing its projected sales in comparison to the actual sales of its best-selling rod-and-reel combinations. Find each difference to show how actual sales compare with projected sales.

Product	Sales Projection	Actual Sales	Difference
Sterling Pro 5	$3,187.15	$3,269.09	_____
Daiwa Pro Flex	$3,724.41	$3,794.35	_____
Cardinal Custom Lite	$2,146.63	$2,226.58	_____
Abu Garcia Custom Lite	$3,129.52	$3,194.46	_____
Eagle Claw Big Water	$3,841.21	$3,881.20	_____
Shimano TX/All Pro	$2,872.33	$2,919.21	_____

HAL'S FISH AND TACKLE

24. What is the difference between the dollar amount of trade with Mexico in 1993 and 1980?

Answer

24. _____

Trade with Mexico has nearly tripled since 1980. Exports to and imports from Mexico, in billions:

$81.5

$27.9

SKILLBUILDER 3.4

Multiplying Decimals

Learning Outcomes

After completing this skillbuilder, you should be able to:

■ Find the product of two decimals.

Multiplying Decimals

Multiply decimals the same way whole numbers are multiplied. When the final product is obtained, place the decimal point so that there are as many decimal places in the answer as there are in the multiplier and multiplicand combined. If there are not enough places in the product, add zeros to the *left* of the product to make up the required number of places.

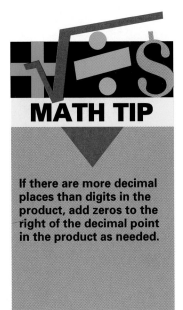

MATH TIP

If there are more decimal places than digits in the product, add zeros to the right of the decimal point in the product as needed.

EXAMPLE

Multiply: 5.089×0.005.

SOLUTION

$$
\begin{array}{rl}
5.089 & \text{3 places} \\
\times\, 0.005 & +\ \text{3 places} \\
\hline
0.025445 & \text{Annex one zero.} \leftarrow \text{6 places}
\end{array}
$$

 **Self-Check**

Multiply: 23.06×0.25.

Self-Check Answer

5.765

Problems

Find the product.

1.   81.7
    × 0.57

2.   0.83
    × 0.08

3.   0.936
    × 7.381

4. $7.15
    ×   65

5. $42.25
    ×   3.5

6. $19.73
    ×   4.3

7. 38.1×0.45

8. 61.7×3.14

9. 6.9×100

10. 75.5×0.16

11.   5.71
    × 10

12. 412.42
    ×   100

13. 7.1956
    × 1,000

14. 86.3×10

15. 41.25×3.8

1. _____
2. _____
3. _____
4. _____
5. _____
6. _____
7. _____
8. _____
9. _____
10. _____
11. _____
12. _____
13. _____
14. _____
15. _____

Problems

16. 52.751×100 **17.** $4.3289 \times 1{,}000$

Calculator Tip

The calculator may not show all the zeros in a product. That is, $23.46 × 5 may appear in the display as 117.3. To write the answer, write a dollar sign and annex a zero: $117.30.

18.	$58.75	**19.**	$61.87	**20.**	$129.52
	× 4.7		× 6.23		× 5.231

Answers

16. _____

17. _____

18. _____

19. _____

20. _____

Business Applications

21. The law firm of Anderson and Anderson employs four office personnel at an average yearly cost of $19,810 each. In order to lower office expenses, the firm has decided to purchase a word processor. With a word processor, the firm will need only two office employees, receiving an average salary of $16,310. Other employee expenses are benefits per employee, $3,500. Additional expenses with the word processor are as follows: information processing equipment, $11,495; workstation, $400; and supplies, $600. Answer the following questions.

a. Find the cost for the year with the word processor.

b. What is the yearly savings by making the change?

Answers

21. a. _____

21. b. _____

22. Arlene works for the State Department. During a recent business trip, Arlene made the following gasoline purchases: 14 gal at $1.479/gal; 18.6 gal at $1.397/gal; 19.5 gal at $1.413/gal; and 16.5 gal at $1.329/gal. How much did Arlene spend for gasoline on her trip? Round each purchase to the nearest cent, where necessary.

Answer

22. _____

23. José, the purchasing agent for Pools Etc., bought pool supplies for the business. Figure the total price of each item, the sales tax (0.06 of the subtotal), and the total cost if there is a $6 shipping fee.

Quantity	Item Number	Description	Unit Price	Total
1	82-050	Test kit	$29.99	_____
15	14-718	Oxidizer	$34.99	_____
7	14-341	Algae control	$14.99	_____
5	14-044	Alkalinity increase	$19.99	_____
5	14-045	pH increase	$24.99	_____
			Subtotal	_____
			Sales tax (0.06)	_____
			Shipping	_____
			Total	_____

Problems

Rounding Quotients

When a number does not divide evenly, you must decide how far to carry out the computation. This depends on the degree of accuracy required. Money amounts are usually rounded to the nearest cent. Divide one place further than the place required, and then round to that place.

EXAMPLE

Divide $49.321 \div 0.121$. Round the quotient to the nearest hundredth.

SOLUTION

```
          407.611  → 407.61
0.121.)49.321.000
       48 4
          921
          847
           74 0
           72 6
            1 40
            1 21
              190
              121
               69
```

If you check a rounded answer using multiplication, the result will not exactly equal the dividend. For example, $407.61 \times 0.121 = 49.32081$. If you round 49.32081 to the nearest thousandth, the result is the dividend, 49.321.

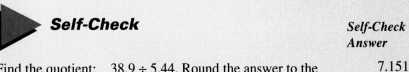 **Self-Check**

Find the quotient: $38.9 \div 5.44$. Round the answer to the nearest thousandth.

Self-Check
Answer

_____7.151_____

Problems

Find the quotient. Round your answers to the nearest hundredth or the nearest cent.

Answers

11. $5.9\overline{)8,148.88}$

12. $64\overline{)\$728.50}$

13. $916\overline{)\$8,914.8}$

14. $694\overline{)\$5,625.97}$

15. $33.60 \div 2.1$

16. $374.4 \div 46$

17. $6.345 \div 3.9$

18. $88.91 \div 712$

11. _____

12. _____

13. _____

14. _____

15. _____

16. _____

17. _____

18. _____

Calculator Tip

Key the division problem 45 ÷ 5.3 into the calculator as follows: 45 (Dividend) ÷ 5.3 (Divisor) =.

Business Applications

19. Schlegel, Markham, and Johnson law firm averages the following fees for these types of cases. The firm charges $90.00 per hour. Find the number of hours spent on the average case. Round your answer to the nearest tenth.

Case Type	Fees Received	No. of Hours
a. Divorce	$ 1,600.00	_____
b. Bankruptcy	1,500.00	_____
c. Real estate	13,000.00	_____
d. Will preparation	400.00	_____

Business Applications

20. The supervisor of trucking at the Crisp Chip Company keeps a record of the fuel mileage for all trucks. Find the miles per gallon (to the nearest tenth) for these trucks in the fleet at Crisp Chip Company.

Truck	Miles	Gallons of Fuel	Miles per Gallon
A	600.0	85.7	_____
B	1,523.0	300.7	_____
C	834.5	92.6	_____
D	643.7	91.95	_____
E	2,542.6	423.766	_____

Business Applications

21. When a new shipment of shoes is delivered to Breymeyer Sporting Goods, the clerk in the athletics shoe department is required to compute and record the cost per pair of shoes. Find the following cost per item (to the nearest cent).

Quantity	Item	Total Cost	Cost per Item
a. 15	Running shoes	$1,039.85	_____
b. 12	Basketball shoes	1,149.49	_____
c. 20	Hiking boots	1,093.85	_____
d. 6	Baseball cleats	283.97	_____
e. 16	Wrestling shoes	517.64	_____
f. 8	Sandals	417.22	_____

Student Notes

SKILLBUILDER 3.6

Shortcuts in Multiplication of Decimals

Learning Outcomes

After completing this skillbuilder, you should be able to:

- Use shortcuts to multiply decimals by 10, 100, or 1,000.
- Use shortcuts to multiply decimals by multiples of 10.
- Use shortcuts to multiply decimals by 25, 50, 125, 250, 500, or 750.

Multiplying by 10, 100, and 1,000

To multiply by 10, move the decimal point of the multiplicand one place to the right. Add a zero if necessary.

To multiply by 100, move the decimal point two places to the right. Add zeros as necessary.

To multiply by 1,000, move the decimal point three places to the right. Add zeros as necessary.

MATH TIP

When multiplying by powers of 10 (such as 10, 100, or 1,000), it is easy to do the calculations mentally by moving the decimal point of the multiplicand one, two, or three places to the right.

EXAMPLE

Use the shortcut to find (a) 495×10, (b) 149.6×100, and (c) $150 \times 1,000$.

SOLUTION

a. $495 \times 10 = 495.0$ or $4,950$

b. $149.6 \times 100 = 149.60$ or $14,960$

c. $150 \times 1,000 = 150.000$ or $150,000$

 **Self-Check**

Use the shortcut to find $47.02 \times 1,000$.

Self-Check Answer

_____47,020_____

Problems

Use the shortcut described on page 113 to find each product. Then total each group.

Answers

1. $ 875 ×   10 =

36.85 × 100 =

8.16 × 1,000 =

62.50 × 100 =

7.95 × 1,000 =

679 ×   10 =

Total =

1. _____

_____

_____

_____

_____

_____

_____

2.   2.36 ×   10 =

18.1 × 100 =

9.74 × 1,000 =

0.6621 × 1,000 =

2.013 ×   10 =

12.9 × 100 =

Total =

2. _____

_____

_____

_____

_____

_____

_____

Multiplying by Multiples of 10

To multiply by a *multiple* of 10 (for example, 40, 300, or 8,000), multiply by the first digit (4, 3, or 8). Then move the decimal point of the answer the required number of places to the right.

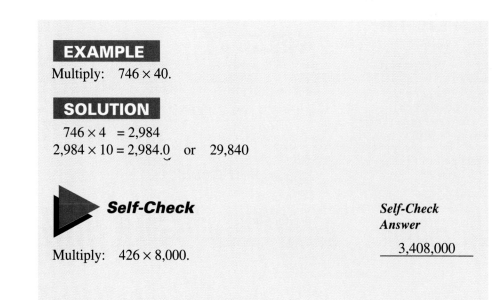

EXAMPLE

Multiply: 746 × 40.

SOLUTION

746 × 4 = 2,984

2,984 × 10 = 2,984.0   or   29,840

Self-Check

Multiply: 426 × 8,000.

Self-Check Answer

3,408,000

Problems

Use the shortcut method described to find each product. Then total each group.

Answers

3. 49 × 50 =

    127 × 80 =

    436 × 300 =

    62 × 9,000 =

    919 × 300 =

    1,585 × 500 =

    Total =

3. _____

_____

_____

_____

_____

_____

_____

4. 327.68 × 400 =

    560.11 × 60 =

    1.2222 × 4,000 =

    500 × 50 =

    616 × 700 =

    45 × 9,000 =

    Total =

4. _____

_____

_____

_____

_____

_____

_____

5. $3,857 × 50 =

    962 × 300 =

    75 × 8,000 =

    432 × 600 =

    56 × 3,000 =

    1,008 × 70 =

    Total =

5. _____

_____

_____

_____

_____

_____

_____

Calculator Tip

Set the decimal selector on F or FL when the maximum number of decimals is needed in the product.

Multiplying by 25 or by a Multiple of 25

To multiply a number by 25, first multiply by 100, then divide by 4.

To multiply by 50, multiply by 100 and divide by 2.

To multiply by 125, first multiply by 1,000 and then divide by 8.

To multiply by 250, multiply by 1,000 and divide by 4. To multiply by 500, multiply by 1,000 and divide by 2. To multiply by 750, multiply by 1,000 and divide by 4; multiply the result by 3.

Problems

EXAMPLE

Multiply: 862×25 and 963×125.

SOLUTION

$862 \times 100 = 862.00$

$86,200 \div 4 = 21,550$

$963 \times 1,000 = 963.000$

$963,000 \div 8 = 120,375$

 **Self-Check**

Multiply: 862×50.

Self-Check Answer

____43,100____

Find the product in each of the following problems using the methods described.

	a.	b.	c.	*Answers*
6.	465×50	785×25	924×25	**6. a.** _____ **b.** _____ **c.** _____
7.	550×125	467×500	73×250	**7. a.** _____ **b.** _____ **c.** _____
8.	675.23×25	943×50	150.50×250	**8. a.** _____ **b.** _____ **c.** _____
9.	179.45×50	65.75×25	$2,026 \times 125$	**9. a.** _____ **b.** _____ **c.** _____
10.	972.87×125	45.38×500	$4,658 \times 250$	**10. a.** _____ **b.** _____ **c.** _____
11.	$6,307 \times 500$	76×125	250×250	**11. a.** _____ **b.** _____ **c.** _____
12.	375×50	$1,280 \times 125$	470×500	**12. a.** _____ **b.** _____ **c.** _____

Business Applications

13. Dr. Sun Li will be attending a medical seminar in Boston, Massachusetts. She is figuring her gas budget for the 750-mile trip. Her car averages 25 mi/gal. If gas prices average 133.9¢ a gallon, how much will the gas for this trip cost?

Answers

13. _____

14. Heath Chip Company has been asked to deliver 100 cases of chips to Louise's Catering Service. Each case sells for $5.95. What is the total amount of the invoice presented to the manager at the catering service if 0.06 of the bill is added for sales tax?

14. _____

15. In preparation for the Kansas Bar Association's Annual Conference, Mr. Cable called the Palace Inn to reserve 750 rooms for three nights. If each room rents for $97.50 a day, what will be the total bill for the conference?

15. _____

Business Applications

16. If the fuel had cost 7.5 cents less per gallon, how much would American Airlines have saved? (Hint: Each cent decrease saves $2.5 million)

16. _____

Pinching pennies on fuel

Jet fuel cost American Airlines 3 cents less per gallon last year than in 1992. What that saved American per month vs. what a 3-cent drop in gasoline prices would have saved you:

American Airlines **$7.5 million**

Car owner **$1.89**

Source: American Airlines, American Automobile Association

By Julie Stacey, USA TODAY

Student Notes

_____

_____

_____

_____

_____

_____

_____

_____

_____

_____

_____

_____

_____

_____

_____

_____

_____

_____

SKILLBUILDER 3.7

Shortcuts in Division of Decimals

Learning Outcomes

After completing this skillbuilder, you should be able to:

- Use shortcuts to divide by multiples of 10.
- Use shortcuts to divide by 25 or multiples of 25.

Dividing by Multiples of 10

To divide by 10, move the decimal point one place to the left. (The decimal point in a whole number is always to the right of the last digit.) For example, $47.693 \div 10 = 4.7693$.

To divide by 100, move the decimal point two places to the left. To divide by 1,000, move the decimal point three places to the left. If necessary, add zeros to the left of the dividend.

If the divisor is a multiple of 10, such as 40, 700, or 3,000, first move the decimal point in the dividend as many places to the left as there are zeros in the divisor. Then divide the resulting number by the nonzero digit (in this case, 4, 7, or 3) to find the quotient.

MATH TIP

Dividing a decimal by a power of 10 can usually be done mentally. You may need a pencil and paper to use some of the other shortcuts, but the work will be easier and quicker than the original problem.

EXAMPLE

Divide: **(a)** $11.5894 \div 1,000$

(b) $96.3 \div 3,000$.

SOLUTION

a. $11.5894 \div 1,000 = 0011.5894 \rightarrow 0.0115894$

b. $96.3 \div 1,000 = 0096.3$

$0.0963 \div 3 = 0.0321$

 **Self-Check**

Divide: **(a)** $4,657 \div 100$ and, **(b)** $1,850 \div 40$.

Self-Check Answer

a. 46.57

b. 46.25

Problems

For each problem, find the quotients and the total. Divide as far as possible. Use the shortcut methods described, and show only your answers. Be sure to align the quotients so that the numbers can be added.

Answers

1. $854 \div 1,000 =$

     $1,467 \div 10 =$

     $42,345 \div 100 =$

     $8,640 \div 500 =$

     $762 \div 40 =$

     $1,000 \div 10 =$

     $10,000 \div 100 =$

     Total $=$

1. _____

_____

_____

_____

_____

_____

_____

_____

2. $19,450 \div 4,000 =$

     $58,209 \div 3,000 =$

     $7,412 \div 800 =$

     $164 \div 400 =$

     $882 \div 700 =$

     $6,500 \div 1,000 =$

     $329 \div 10 =$

     Total $=$

2. _____

_____

_____

_____

_____

_____

_____

_____

3. $\$757.42 \div 80 =$

     $96.844 \div 400 =$

     $357.64 \div 2,000 =$

     $327.70 \div 500 =$

     $900.00 \div 10 =$

     $65.92 \div 4,000 =$

     $947.65 \div 100 =$

     Total $=$

3. _____

_____

_____

_____

_____

_____

_____

_____

Calculator Tip

When working with numbers that have a large number of places, your calculator may not be able to display all the places. Shortcuts are useful for getting exact answers in such cases.

Dividing by 25 or a Multiple of 25

To divide by 25, first divide by 100. Then multiply the result by 4. This works because dividing by 25 gives a quotient four times larger than dividing by 100.

To divide by 50, first divide by 100. Then multiply the result by 2.

To divide by 125, first divide by 1,000. Then multiply the result by 8.

To divide by 250, first divide by 1,000. Then multiply the result by 4. To divide by 500, first divide by 1,000. Then multiply the result by 2.

Problems

EXAMPLE

Divide: (a) $429 \div 25$ and (b) $\$6,160 \div 125$.

SOLUTION

a. $429 \div 100 = 429.$
$4.29 \times 4 = 17.16$

b. $\$6,160 \div 1,000 = \$6160.$
$\$6.16 \times 8 = \$49.28$

▶ **Self-Check**

Divide: $357 \div 50$.

Self-Check Answer

_____7.14_____

Find the quotient in each of the following problems. Use the shortcut methods, and show your work. For money amounts, round your final answers to the nearest cent where necessary.

Answers

4. $3,260 \div 25$ **5.** $\$1,729.50 \div 50$ **6.** $8,475 \div 250$

4. _____

5. _____

6. _____

7. _____

7. $\$6,750 \div 125$ **8.** $\$418.90 \div 125$ **9.** $\$202.50 \div 25$

8. _____

9. _____

10. _____

10. $4,650 \div 125$ **11.** $217.6 \div 250$ **12.** $9,850 \div 500$

11. _____

12. _____

13. _____

14. _____

13. $13,500 \div 25$ **14.** $33,750 \div 500$ **15.** $847.5 \div 50$

15. _____

16. _____

17. _____

16. $\$17,295 \div 250$ **17.** $\$21,760 \div 125$ **18.** $4,930 \div 500$

18. _____

Business Applications

19. The State Department of Transportation needs to decide which state vehicle to send on long trips. Each employee has been asked to report the miles per gallon for his or her issued car. Robert Parker made a 395.5-mi trip on 10 gallons of gas. How many miles per gallon should Robert report for his car?

19. _____

20. Maria Morian, attorney-at-law, rendered services at a total cost of $1,875 to a client in bankruptcy court. Her hourly fee for such services is $125.00. The bill sent to her client stated the total number of hours worked on the case. For how many hours was the client billed?

20. _____

21. Brandon Brittain is restocking the shelves in his baseball card shop. He purchased 375 boxes of baseball cards. If each crate holds 25 boxes, how many crates did he purchase for his shop?

21. _____

Business Applications

22. Compute the average decrease in Japanese imports from 1988 to 1993.

Answers

22. _____

_____

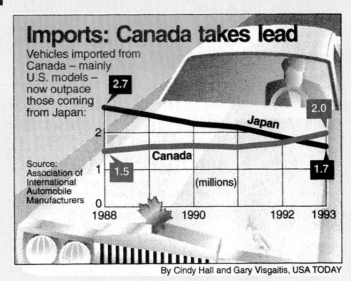

Imports: Canada takes lead

Vehicles imported from Canada – mainly U.S. models – now outpace those coming from Japan:

2.7

2.0

Japan

Canada

Source: Association of International Automobile Manufacturers

(millions)

1.5

1.7

1988   1990   1992 1993

By Cindy Hall and Gary Visgaitis, USA TODAY

Student Notes

_____

_____

_____

_____

_____

_____

_____

_____

_____

_____

_____

_____

_____

_____

_____

_____

U N I T

4

WORKING WITH FRACTIONS

Fractions are often used when discussing data, especially in the stock market. As the stock market goes up and down, you will need to be able to add and subtract fractions in order to keep track of your company's stock prices. Which of the stocks shown pay dividends? How many increased in price?

In this unit we study how to add, subtract, multiply, and divide fractions, as well as learn shortcuts for computing with common fractional parts.

NEW YORK (A9) - - Following are complete yearly nationwide composite prices for stocks listed on the New York Stock Exchange.

-01- YRLY-NYSE for JAN 01

Name	Div	Sales (hds)	365-Day High	Low	Close	Chg.
			- A - A -			
AAR	.48	81619	15	$11\frac{1}{2}$	$14\frac{1}{2}$	+ $2\frac{3}{4}$
ACE Lt n	.43	349641	36	$25\frac{1}{2}$	$31\frac{1}{8}$	+ $2\frac{1}{2}$
ACMin	1.10	243334	$12\frac{5}{8}$	$10\frac{3}{4}$	$12\frac{1}{4}$	+ $1\frac{1}{4}$
ACM Op	.80	55391	$10\frac{1}{4}$	$9\frac{1}{8}$	$9\frac{3}{4}$	+ $\frac{3}{8}$
ACM Sc	1.09	355534	$12\frac{1}{4}$	$10\frac{1}{2}$	$11\frac{7}{8}$	+ $1\frac{1}{4}$
ACMSp	.96	202298	$10\frac{5}{8}$	$8\frac{3}{4}$	10	+ 1
ACMMD n	.12	32838	$15\frac{1}{8}$	14	$14\frac{1}{2}$	− $\frac{1}{2}$
ACM MI	1.08	102899	$12\frac{3}{8}$	$9\frac{5}{8}$	$11\frac{1}{4}$	+ $1\frac{1}{8}$
ACMMM	.72	54633	$9\frac{3}{8}$	$8\frac{5}{8}$	9	+ $\frac{1}{4}$
ACMMu n	.90	35250	$15\frac{1}{8}$	$12\frac{7}{8}$	$13\frac{1}{2}$	− $1\frac{1}{2}$
ADT wt		110073	2	$\frac{3}{4}$	$\frac{13}{16}$	− $\frac{5}{16}$
ADT		610714	$10\frac{3}{8}$	$6\frac{7}{8}$	$8\frac{1}{8}$	+ $1\frac{1}{8}$
AFLAC s	.40	405255	34	$24\frac{3}{4}$	$28\frac{1}{2}$	+ $\frac{7}{8}$
AL Lab	.18	183508	$29\frac{3}{8}$	$12\frac{3}{4}$	14	−$11\frac{7}{8}$
AMR		1570478	$72\frac{7}{8}$	$55\frac{1}{2}$	67	− $\frac{1}{2}$
ANR pf	2.67	217	$27\frac{7}{8}$	$25\frac{3}{4}$	$25\frac{3}{4}$	− $\frac{7}{8}$
ANR pfB	2.12	752	27	25	$25\frac{1}{4}$	+ $\frac{1}{8}$
ARCOCh	2.50	74079	$47\frac{1}{4}$	$39\frac{1}{4}$	$43\frac{1}{4}$	− $\frac{1}{2}$
ARX		62267	$3\frac{7}{8}$	$1\frac{3}{4}$	$3\frac{1}{2}$	+ $1\frac{1}{2}$

Financial Planner

Financial planners often invest money for their clients, helping them to plan for retirement. Most financial planners will invest in either individual stocks or mutual funds. However, to ensure that a client does not sustain a devastating loss, planners keep track of stocks constantly. The New York Stock Exchange lists each stock and gives important information such as the volume of trading (stocks bought and sold). The last four columns list the highest price, lowest price, closing price, and the change from the prior day's closing price. Financial planners will watch for a trend and buy or sell when appropriate. It is interesting to note that the change is in fractions, not decimals.

Math Application

Three years ago, you bought 15 shares of Company ABC, at a price of $28\frac{1}{2}$ per share. Your financial planner calls this morning to let you know that your stock's closing price yesterday was $29\frac{1}{8}$. If you could sell your stock for $29\frac{1}{8}$ per share, how much would you gain?

There are actually two ways to compute the gain.

1. Compute the total value of the stock at each price and subtract.

$$(15)\,(29\tfrac{1}{8}) - (15)\,(28\tfrac{1}{2}) = \$ \rule{2cm}{0.4pt}$$

2. Find the gain on each share and then multiply by the number of shares to get the total gain.

$$(29\tfrac{1}{8} - 28\tfrac{1}{2})\,(15) = \$ \rule{2cm}{0.4pt}$$

Critical Thinking Problem

Some stocks require that a fee be paid up front in order to make a purchase. These stocks may be more stable or may outperform other stocks over time. What issues might be pertinent to your decision about which stocks to buy?

SKILLBUILDER 4.1

Renaming Fractions

Learning Outcomes

After completing this skillbuilder, you should be able to:

- Rename fractions in lower or higher terms.
- Rename improper fractions as whole or mixed numbers and vice versa.
- Rename fractions as decimals and vice versa.

Renaming Fractions

A fraction is a number written in this form: $\frac{2}{3}$. The number above the line is the **numerator;** the number below the line is the **denominator.** The numerator and denominator are the **terms** of the fraction. A fraction whose numerator is smaller than its denominator is a **proper fraction** and always has a value of less than 1.

To rename a fraction in lower terms, find a number that is contained an even number of times in both the numerator and the denominator. Dividing both the numerator and the denominator by this number will result in a fraction equal in value to the original fraction, but made up of smaller numbers. When no number other than 1 will divide evenly into both terms, a fraction is said to be **in lowest terms.** Find the greatest number that will divide both the numerator and denominator evenly. Then divide both the numerator and denominator by that number.

When a fraction must be renamed as an equivalent fraction with a larger denominator, divide the new denominator (the larger term) by the denominator of the fraction being renamed. Use this quotient to multiply both the numerator and denominator of the fraction to be renamed.

Problems

MATH TIP

When you multiply or divide both the numerator and denominator by the same number, it is the same as multiplying the number by 1. The value of a number does not change when you divide or multiply it by 1.

EXAMPLE

a. Write $\frac{12}{28}$ in lowest terms.

b. Rename $\frac{2}{15}$ with a denominator of 60.

SOLUTION

a. $\frac{12}{28} = \frac{12 \div 4}{28 \div 4} = \frac{3}{7}$

Note that no number other than 1 divides both 3 and 7 evenly, so $\frac{3}{7}$ is in lowest terms.

b. $60 \div 15 = 4$

$$\frac{2}{15} = \frac{2 \times 4}{15 \times 4} = \frac{8}{60}$$

Because $\frac{4}{4} = 1$, multiplying or dividing any fraction by $\frac{4}{4}$ does not change the real value of the fraction.

 **Self-Check**

a. Write $\frac{15}{35}$ in lowest terms.

b. Rename $\frac{3}{4}$ with a denominator of 20.

Self-Check Answers

$\frac{3}{7}$

_____

$\frac{15}{20}$

_____

Rename the following fractions in lowest terms.

Answers

1. $\frac{6}{16}$ **2.** $\frac{15}{35}$ **3.** $\frac{16}{24}$

4. $\frac{27}{81}$ **5.** $\frac{18}{72}$ **6.** $\frac{9}{45}$

7. $\frac{14}{20}$ **8.** $\frac{36}{42}$ **9.** $\frac{27}{33}$

1. _____

2. _____

3. _____

4. _____

5. _____

6. _____

7. _____

8. _____

9. _____

Problems

10. $\dfrac{18}{27}$ **11.** $\dfrac{14}{84}$ **12.** $\dfrac{21}{42}$

10. _____

11. _____

12. _____

13. $\dfrac{14}{42}$ **14.** $\dfrac{26}{39}$ **15.** $\dfrac{15}{25}$

13. _____

14. _____

15. _____

16. $\dfrac{28}{40}$ **17.** $\dfrac{17}{51}$ **18.** $\dfrac{28}{36}$

16. _____

17. _____

18. _____

Rename the following fractions with the indicated denominator.

19. $\dfrac{2}{3}, 27$ **20.** $\dfrac{3}{8}, 32$ **21.** $\dfrac{2}{5}, 40$ **22.** $\dfrac{11}{12}, 36$

19. _____

20. _____

21. _____

22. _____

23. $\dfrac{4}{32}, 96$ **24.** $\dfrac{5}{18}, 72$ **25.** $\dfrac{3}{7}, 56$ **26.** $\dfrac{4}{15}, 90$

23. _____

24. _____

25. _____

26. _____

Renaming Fractions as Mixed Numbers and Vice Versa

A fraction whose numerator is equal to or greater than its denominator is an **improper fraction.** A **mixed number** is a combination of a whole number and a fraction. To rename an improper fraction as a mixed number, divide the denominator into the numerator. The quotient is the whole-number part. The fractional part is formed by the remainder (which becomes the numerator) and the denominator of the original fraction. The fraction should be renamed in lowest terms.

To rename a whole number as a fraction, write the whole number as the numerator with a denominator of 1. To rename a mixed number as an improper fraction, multiply the whole number by the denominator in the fraction, and add the numerator of the fraction to this product. Place this new numerator over the denominator of the mixed number fraction.

Problems

EXAMPLE

a. Rename $\frac{30}{4}$ as a mixed number.

b. Rename 15 as an improper fraction.

c. Rename $7\frac{2}{5}$ as an improper fraction.

SOLUTION

a. $\frac{30}{4} = 7\frac{2}{4} = 7\frac{1}{2}$

b. $15 = \frac{15}{1}$

c. $7\frac{2}{5}$ $\qquad$ $7 \times 5 + 2 = 37$

$\qquad\qquad$ $7\frac{2}{5} = \frac{37}{5}$

▶ Self-Check

a. Rename $\frac{29}{4}$ as a mixed number.

b. Rename $8\frac{5}{7}$ as an improper fraction.

Self-Check Answer

a. $7\frac{1}{4}$ _____

b. $\frac{61}{7}$ _____

Rename the following improper fractions as whole or mixed numbers.

Answers

27. $\frac{12}{10}$ $\qquad$ **28.** $\frac{9}{9}$ $\qquad$ **29.** $\frac{58}{26}$

27. _____

28. _____

29. _____

30. $\frac{117}{20}$ $\qquad$ **31.** $\frac{51}{12}$ $\qquad$ **32.** $\frac{46}{14}$

30. _____

31. _____

32. _____

Rename these whole numbers as improper fractions.

33. 75 $\qquad$ **34.** 13 $\qquad$ **35.** 4 $\qquad$ **36.** 166 $\qquad$ **37.** 1

33. _____

34. _____

35. _____

36. _____

37. _____

Rename the following mixed numbers as improper fractions.

38. $5\frac{4}{6}$ $\qquad$ **39.** $16\frac{2}{3}$ $\qquad$ **40.** $3\frac{5}{9}$

38. _____

39. _____

40. _____

Problems

41. $3\frac{5}{7}$ **42.** $4\frac{5}{11}$ **43.** $12\frac{3}{5}$

41. _____

42. _____

43. _____

Renaming Fractions as Decimals and Vice Versa

To rename a decimal fraction as a proper fraction, write the decimal fraction as the numerator (without a decimal point) over a denominator of 1, plus as many zeros as there are places to the right of the decimal point.

To rename a proper fraction as a decimal fraction, divide the numerator by the denominator. When the division ends in a run-on number, either round the run-on number or express it as a common fraction.

Calculator Tip

When using a calculator, round repeating decimals to five places by setting 5 as the desired number of places.

EXAMPLE

Rename 0.56 and $0.5\frac{1}{2}$ as proper fractions in lowest terms.

SOLUTION

$$0.56 = \frac{56}{100} = \frac{56 \div 4}{100 \div 4} = \frac{14}{25}$$

$$0.5\frac{1}{2} = 0.55 = \frac{55}{100} = \frac{55 \div 5}{100 \div 5} = \frac{11}{20}$$

EXAMPLE

Rename $\frac{3}{8}$ and $\frac{1}{6}$ as decimals.

SOLUTION

$$\begin{array}{r} 0.375 \\ 8\overline{)3.000} \end{array} \qquad \frac{3}{8} = 0.375$$

$$\begin{array}{r} 0.1666 \\ 6\overline{)1.0000} \end{array} \qquad \frac{1}{6} = 0.16\frac{2}{3}, \text{ or } 0.167$$

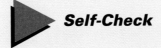 **Self-Check**

a. Rename $\frac{5}{12}$ as a decimal.

b. Rename 0.78 as a proper fraction in lowest terms.

Self-Check Answers

a. $0.41\frac{2}{3}$, or 0.417

b. $\frac{39}{50}$

Problems

Rename the following fractions as decimals. Round to the third decimal place if necessary.

44. $\dfrac{5}{16}$ **45.** $\dfrac{4}{32}$ **46.** $\dfrac{1}{9}$ **47.** $\dfrac{28}{64}$

44. _____

45. _____

46. _____

47. _____

Rename the following decimal fractions as proper fractions. Rename the answers in lowest terms.

48. 0.48   **49.** 0.625   **50.** 0.375

48. _____

49. _____

50. _____

51. _____

51. $0.16\dfrac{1}{4}$ **52.** $0.22\dfrac{2}{5}$ **53.** $0.11\dfrac{1}{9}$

52. _____

53. _____

54. _____

54. 0.1875   **55.** 0.15625   **56.** 1.80

55. _____

56. _____

57. _____

58. _____

57. 2.125   **58.** $3.366\dfrac{2}{3}$ **59.** $0.2727\dfrac{3}{11}$

59. _____

Business Applications

60. The president of EDA Brokerage keeps careful track of the stock market. The following stock market summary lists stock measures in decimal form. Change each to a mixed number in lowest terms.

DOW JONES AVERAGES			
20	Utilities	341.49	_____
35	Transportation	1,425.32	_____
70	Stocks	825.21	_____
20	Bonds	94.18	_____
10	Public utility bonds	83.85	_____
20	Industrial bonds	80.03	_____
40	Industrials	2,543.07	_____
	Commodity futures	135.69	_____

61. Bikes and More sold 500 bikes in June. Of these, 225 were Spirits, 178 were Hill-Climbers, 75 were Easy Riders, and 22 were special-order racing bikes. Find the fraction (in lowest terms) of each type of bike sold.

Answers

61. _____

_____

_____

_____

62. Chambers Corp. invested in several stocks. On a recent day, they listed the quoted prices of the stocks, which are given in dollars and fractions of a dollar. Write the stock prices in dollars and cents. Round to the nearest cent if necessary.

Answers

62. _____

_____

_____

_____

_____

Stock	Price
PepsiCo	$45\frac{1}{4}$
Wal-Mart	$52\frac{1}{2}$
IBM	$22\frac{7}{8}$
Ford	$26\frac{3}{8}$
Colgate/Palmolive	$39\frac{1}{2}$

63. Write each batting average as a fraction, reducing to lowest terms.

63. _____

_____

_____

_____

Best 1993 averages for at-bats when they were behind in the count

Gregg Jefferies St. Louis	.329
Barry Bonds San Francisco	.303
Ryne Sandberg Chicago Cubs	.298
Alan Trammell Detroit	.296

Tomorrow: Pitching out of a jam

Source: STATS 1994 Player Profiles   By John Riley and Sam Ward, USA TODAY

SKILLBUILDER **4.2**

Finding the Lowest Common Denominator

Learning Outcomes

After completing this skillbuilder, you should be able to:

■ Find the lowest common denominator of a group of fractions.

Finding the Lowest Common Denominator

When two or more fractions are to be added or subtracted, all the fractions must have the same denominator. If the denominators are not the same, the fractions must be renamed as fractions with the same denominator. The smallest denominator which is common to all the fractions—the lowest common denominator (LCD)—should be used. One of these methods can be used to find the LCD.

1. If the smaller denominator of a pair of fractions divides evenly into the larger denominator, the larger denominator is the LCD.
2. Multiply the larger denominator by 2. Check to see if the product is divisible by the smaller denominator. If it is not, multiply the larger denominator by 3, 4, and so on, until you find the smallest product divisible evenly by the smaller denominator.

    If the lowest common denominator is not obvious by the previous methods, use the prime number method.

3. *Prime numbers.* A **prime number** is one that is divisible only by 1 and itself. The numbers 2, 3, 5, 7, 11, 13, and 17 are examples of prime numbers. To determine the lowest common denominator by this method, do the following steps.

    a. List all the denominators.
    b. Beginning with the lowest prime number other than 1 as the divisor, divide the denominators and bring down the results.
    c. Continue dividing the results by prime numbers until all the results are reduced to 1.
    d. Find the product of all the prime numbers used. The product is the lowest common denominator.

Problems

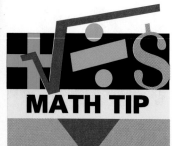

MATH TIP

When possible, determine the LCD mentally.

EXAMPLE

1. Find the LCD of $\frac{1}{3}$ and $\frac{7}{12}$.

2. Find the LCD of $\frac{8}{15}$ and $\frac{11}{25}$.

3. Use prime numbers to find the LCD of $\frac{3}{8}$, $\frac{1}{7}$, $\frac{7}{10}$, $\frac{5}{16}$, and $\frac{2}{5}$.

SOLUTION

1. $12 \div 3 = 4$, so the smaller denominator evenly divides the larger one. The LCD is 12.

2. 15 does not divide 25 evenly.

   $25 \times 2 = 50$ 15 does not divide 50 evenly.

   $25 \times 3 = 75$ $75 \div 15 = 5$

   The LCD is 75.

3.
a 2⌐8	7	10	16	5	Because 2 does not divide
b 2⌐4	7	5	8	5	7 and 5 evenly, bring the 7
2⌐2	7	5	4	5	and 5 down unchanged.
2⌐1	7	5	2	5	
5⌐1	7	5	1	5	Because 5 does not divide
c 7⌐1	7	1	1	1	7 evenly, bring the 7 down
1	1	1	1	1	unchanged.

   d $2 \times 2 \times 2 \times 2 \times 5 \times 7 = 560$

              lowest common denominator

▶ Self-Check

Find the LCD: $\frac{2}{3}$, $\frac{3}{5}$, $\frac{7}{8}$, $\frac{1}{2}$.

Self-Check Answer

_____120_____

Find the lowest common denominator for the following groups of fractions.

1. $\frac{1}{2}$, $\frac{2}{3}$, $\frac{3}{4}$, $\frac{1}{6}$

2. $\frac{1}{2}$, $\frac{5}{8}$, $\frac{7}{18}$, $\frac{3}{4}$

3. $\frac{2}{3}$, $\frac{1}{4}$, $\frac{4}{15}$, $\frac{7}{8}$

4. $\frac{1}{9}$, $\frac{3}{8}$, $\frac{5}{12}$, $\frac{1}{4}$

Answers

1. _____

2. _____

3. _____

4. _____

Problems

5. $\dfrac{3}{4}$, $\dfrac{3}{16}$, $\dfrac{2}{3}$, $\dfrac{5}{8}$

6. $\dfrac{3}{5}$, $\dfrac{4}{15}$, $\dfrac{11}{12}$, $\dfrac{2}{3}$

Answers

5. _____

6. _____

7. _____

8. _____

9. _____

10. _____

11. _____

12. _____

13. _____

14. _____

7. $\dfrac{1}{3}$, $\dfrac{5}{21}$, $\dfrac{4}{9}$, $\dfrac{1}{15}$

8. $\dfrac{7}{12}$, $\dfrac{2}{3}$, $\dfrac{1}{8}$, $\dfrac{3}{5}$

9. $\dfrac{2}{3}$, $\dfrac{1}{15}$, $\dfrac{3}{4}$, $\dfrac{5}{8}$, $\dfrac{1}{2}$

10. $\dfrac{1}{2}$, $\dfrac{5}{6}$, $\dfrac{1}{7}$, $\dfrac{1}{4}$, $\dfrac{1}{3}$

11. $\dfrac{5}{12}$, $\dfrac{1}{3}$, $\dfrac{1}{2}$, $\dfrac{5}{9}$, $\dfrac{1}{7}$

12. $\dfrac{3}{5}$, $\dfrac{1}{4}$, $\dfrac{1}{11}$, $\dfrac{7}{15}$, $\dfrac{1}{8}$

Calculator Tip

You can use a calculator to arrive at the LCD by dividing the multiples of the greater denominator by the lesser denominator until you find the LCD.

13. $\dfrac{1}{6}$, $\dfrac{7}{30}$, $\dfrac{3}{16}$, $\dfrac{5}{32}$, $\dfrac{3}{8}$

14. $\dfrac{3}{13}$, $\dfrac{5}{9}$, $\dfrac{7}{12}$, $\dfrac{1}{8}$, $\dfrac{11}{20}$

Business Applications

15. The Brownville Cab Company recorded the hours worked by each employee on Saturday. Find the LCD of the fractions.

Employee	Hours
Kathy Munoz	$7\frac{1}{2}$
Sun Lee	8
Sean O'Connor	$7\frac{3}{4}$
Rob Thompkins	$6\frac{5}{8}$
Karen Mikovec	$5\frac{21}{26}$

16. Frank Jones, attorney-at-law, kept track of the hours he spent on a divorce case. Find the LCD of the fractions.

16. _____

Days	Hours
Monday	$2\frac{5}{16}$
Tuesday	$4\frac{2}{3}$
Wednesday	0
Thursday	$1\frac{1}{2}$
Friday	$3\frac{5}{12}$

Business Applications

17. Javiar Lopez is a mechanic at the Tastee Ice Cream Company Shops. In one week he worked the following hours. Find the LCD of the fractions.

Days	Hours
Sunday	$8\frac{1}{2}$
Monday	$9\frac{3}{8}$
Tuesday	10 h
Wednesday	$9\frac{3}{4}$
Thursday	$10\frac{3}{5}$

18. Compute the lowest common denominator of 3.25, 3.50, and 3.75, for the interest rates listed below.

18. _____

Rate hikes

The fed funds rate had been at 3% since September 1992. This year the Fed pushed it up:

Date	New rate
Feb. 4	3.25%
March 22	3.50%
April 18	3.75%

USA TODAY research

Student Notes

_____

_____

_____

_____

Student Notes

SKILLBUILDER **4.3**

Adding Fractions and Mixed Numbers

Learning Outcomes

After completing this skillbuilder, you should be able to:

- Add fractions or mixed numbers with the same or different denominators.
- Add fractions or mixed numbers by writing the fraction in decimal form.

Adding Fractions and Mixed Numbers

Fractions can be added only when they have like denominators. The total of the numerators of the addends is the numerator of the sum. The common denominator of the addends becomes the denominator of the sum. Rename the answer in lowest terms if necessary.

When the denominators are not all the same, the lowest common denominator must be found; all fractions must then be renamed in like terms.

EXAMPLE

a. Add: $\frac{7}{24} + \frac{11}{24}$.

b. Add: $\frac{4}{5} + \frac{3}{5}$.

c. Add: $27\frac{3}{8} + 13\frac{5}{16} + 3\frac{1}{3} + 2\frac{5}{6}$.

SOLUTION

a.
$$\begin{array}{r} \frac{7}{24} \\ +\frac{11}{24} \\ \hline \frac{18}{24} = \frac{3}{4} \end{array}$$

b.
$$\begin{array}{r} \frac{4}{5} \\ +\frac{3}{5} \\ \hline \frac{7}{5} = 1\frac{2}{5} \end{array}$$

Problems

c. The LCD is 48. Each mixed number is rewritten with a denominator of 48. Then the fractions are added. Finally, the whole numbers are added, and the resulting mixed number is written in lowest terms.

$$27\frac{3}{8} = 27\frac{18}{48}$$

$$13\frac{5}{16} = 13\frac{15}{48}$$

$$3\frac{1}{3} = 3\frac{16}{48}$$

$$2\frac{5}{6} = 2\frac{40}{48}$$

$$45\frac{89}{48} = 45 + 1\frac{41}{48} = 46\frac{41}{48}$$

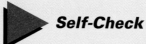 **Self-Check**

Add the following fractions: $6\frac{7}{12} + 2\frac{3}{8} + 5\frac{1}{6}$.
Rename the answer in lowest terms.

Self-Check Answer

$$14\frac{1}{8}$$

Add the following groups of fractions. Rename the answers in lowest terms.

1. $2\frac{1}{4}$

$4\frac{3}{8}$

$3\frac{11}{32}$

$2\frac{1}{2}$

$+\ 3\frac{1}{16}$

2. $7\frac{7}{18}$

$3\frac{5}{6}$

$3\frac{1}{3}$

$5\frac{5}{12}$

$+\ 2\frac{5}{9}$

Answers

1. _____

2. _____

Problems

3. $16\frac{1}{8}$

$9\frac{2}{5}$

$2\frac{7}{15}$

$3\frac{7}{9}$

$+\ \ 1\frac{1}{16}$
_____

4. $4\frac{3}{5}$

$2\frac{3}{7}$

$5\frac{3}{4}$

$2\frac{3}{14}$

$+\ 1\frac{1}{8}$
_____

Answers

3. _____

4. _____

5. _____

6. _____

7. _____

8. _____

9. _____

5. $4\frac{3}{8}$

$3\frac{2}{3}$

$3\frac{7}{12}$

$2\frac{4}{9}$

$+\ 3\frac{1}{2}$
_____

6. $5\frac{7}{30}$

$2\frac{1}{11}$

$3\frac{5}{6}$

$2\frac{2}{3}$

$+\ 6\frac{2}{5}$
_____

7. $4\frac{1}{4}$

$2\frac{1}{3}$

$3\frac{1}{3}$

$9\frac{3}{4}$

$+\ 1\frac{1}{8}$
_____

8. $12\frac{1}{6}$

$9\frac{2}{3}$

$5\frac{5}{6}$

$11\frac{5}{24}$

$+\ 10\frac{1}{8}$
_____

9. $8\frac{5}{8}$

$19\frac{2}{5}$

$6\frac{2}{9}$

$23\frac{5}{12}$

$+\ 10\frac{1}{4}$
_____

Problems

Adding Fractions Using Decimal Equivalents

Calculator Tip

To use a calculator to add fractions, write all fractions in decimal form. You can use the memory to store each fraction as you rewrite it in decimal form and then use memory recall to determine the sum.

Fractions can also be added by using their decimal equivalents. A fraction is expressed as its decimal equivalent by dividing the denominator into the numerator.

Sometimes a group of numbers contains both decimal fractions and common fractions. When this occurs, express all the numbers either as decimal fractions or as common fractions before proceeding with the addition.

EXAMPLE

a. Add in decimal form and fraction form: $6\frac{1}{2} + 3\frac{5}{8} + 4\frac{1}{16} + 2\frac{3}{4}$.

b. Add: $0.37\frac{1}{2} + 0.42\frac{1}{4} + \frac{5}{8} + 0.48\frac{2}{3}$. Round to four decimal places, where necessary.

SOLUTION

a.

Add as Decimals	Add as Fractions
6.5000	$6\frac{1}{2} = 6\frac{8}{16}$
3.6250	$3\frac{5}{8} = 3\frac{10}{16}$
4.0625	$4\frac{1}{16} = 4\frac{1}{16}$
+ 2.7500	$+ 2\frac{3}{4} = 2\frac{12}{16}$
16.9375	$15\frac{31}{16} = 16\frac{15}{16}$

Note: 0.9375 is equivalent to $\frac{15}{16}$.

b.

$0.37\frac{1}{2} = 0.3750$

$0.42\frac{1}{4} = 0.4225$

$\frac{5}{8} = 0.6250$

$+ 0.48\frac{2}{3} = \dfrac{0.4867}{1.9092}$

▶ **Self-Check**

Add: $0.62\frac{1}{2}$, $\frac{1}{7}$, and $3.7\frac{1}{4}$. Round to five decimal places, where necessary.

Self-Check Answer

4.49286

Problems

Add the following groups of numbers by renaming all addends as decimal fractions. Round to five decimal places, where necessary.

10.
$$\frac{5}{16}$$
$$0.37\frac{1}{2}$$
$$0.15\frac{1}{8}$$
$$+\quad\frac{1}{32}$$

11. $7.16\frac{3}{4}$
$6.25\frac{2}{5}$
$8.9\frac{5}{6}$
$+\,6.02\frac{1}{4}$

12. $0.87\frac{1}{2}$
$6.7\frac{1}{4}$
$4.7\frac{1}{2}$
$+\quad\frac{5}{8}$

10. ——————————
11. ——————————
12. ——————————
13. ——————————
14. ——————————
15. ——————————
16. ——————————
17. ——————————
18. ——————————

13. $0.37\frac{3}{4}$
$5.43\frac{5}{6}$
$0.93\frac{1}{5}$
$+\,9\frac{1}{8}$

14.
$$\frac{5}{8}$$
$$0.6\frac{1}{4}$$
$$0.34\frac{3}{8}$$
$$+\,45\frac{1}{4}$$

15. $0.45\frac{1}{2}$
$$\frac{1}{16}$$
$$3.3\frac{1}{4}$$
$$+\,4\frac{3}{16}$$

16. $7.15\frac{3}{4}$
0.94
$2.33\frac{1}{16}$
$+\,4.37\frac{1}{2}$

17. $0.92\frac{1}{16}$
$2.03\frac{1}{2}$
$8.43\frac{3}{32}$
$+\,10\frac{3}{16}$

18.
$$\frac{3}{32}$$
$$0.42\frac{1}{2}$$
$$3.25\frac{2}{5}$$
$$+\,55\frac{1}{4}$$

Business Applications

19. During inventory, Week's Paint Store located the following quantities of paint stored in the stock room: $22\frac{1}{2}$ gal of burnt orange, $23\frac{1}{2}$ gal of sky blue, $34\frac{1}{4}$ gal of royal blue, $11\frac{3}{4}$ gal of bright yellow, and $35\frac{1}{4}$ gal of beige. How much paint is stored in the stock room?

19. _____

20. Luanda Jackson, a plumber, worked $2\frac{1}{4}$ hr of overtime on Tuesday, $1\frac{1}{2}$ hr on Wednesday, $2\frac{2}{3}$ hr on Friday, and $7\frac{3}{4}$ hr on Saturday. Find the total number of overtime hours she worked.

20. _____

21. Change the men's decimal percents to fraction percents and add to get the total % of men responding.

21. _____

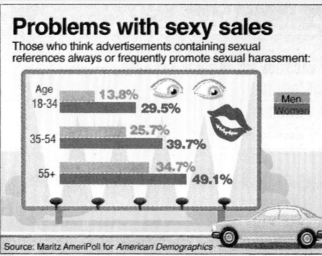

Problems with sexy sales

Those who think advertisements containing sexual references always or frequently promote sexual harassment:

Age	Men	Women
18-34	13.8%	29.5%
35-54	25.7%	39.7%
55+	34.7%	49.1%

Source: Maritz AmeriPoll for American Demographics

By Cindy Hall and Cliff Vancura, USA TODAY

SKILLBUILDER 4.4

Subtracting Fractions and Mixed Numbers

Learning Outcomes

After completing this skillbuilder, you should be able to:

- Find the difference of two fractions or mixed numbers.

Subtracting Fractions and Mixed Numbers

As in addition, like denominators must be used in order to subtract fractions. Subtract the numerator of the subtrahend from the numerator of the minuend. The common denominator of the two fractions will be the denominator of the difference. Rename in lowest terms if necessary.

When mixed numbers are subtracted, the fraction in the subtrahend may be larger than the fraction in the minuend. If so, subtract one unit from the whole number in the minuend. Change it to a fraction and add it to the fraction in the minuend. Once the borrowing is accomplished, subtract the whole numbers and the numerators of the fractions.

Problems

MATH TIP

To reduce fractions, use the rules of divisibility. These include the following: A number is divisible by 2 if it is an even number, 3 if the sum of the digits is divisible by 3, and 5 if the number ends in a 0 or a 5.

EXAMPLE

a. Subtract: $\frac{7}{8} - \frac{3}{4}$.

b. Subtract: $9\frac{3}{8} - 5\frac{5}{8}$.

SOLUTION

a.

$$\frac{7}{8} = \frac{7}{8}$$
$$-\frac{3}{4} = \frac{6}{8}$$
$$\overline{} \quad \frac{1}{8}$$

b. Since $\frac{5}{8}$ is greater than $\frac{3}{8}$, we must rewrite one unit from the whole number, 9, as a fraction and add it to $\frac{3}{8}$.

$$9\frac{3}{8} = 8 + 1 + \frac{3}{8} = 8 + \frac{8}{8} + \frac{3}{8} = 8\frac{11}{8}$$
$$-5\frac{5}{8} = 5\frac{5}{8} \qquad = 5\frac{5}{8} \qquad = 5\frac{5}{8}$$
$$\overline{} \quad 3\frac{6}{8}, \text{ or } 3\frac{3}{4}$$

▶ **Self-Check**

Subtract: $18\frac{5}{6} - 12\frac{15}{16}$.

Self-Check Answer

$$5\frac{43}{48}$$

Find the difference in each problem. Rename in lowest terms, if necessary.

Answers

1. $24\frac{3}{8}$
$-11\frac{5}{6}$

2. $35\frac{1}{12}$
$-13\frac{5}{12}$

3. $7\frac{2}{9}$
$-3\frac{5}{16}$

1. _____

2. _____

3. _____

Problems

Answers

4. $54\frac{2}{5}$ **5.** $36\frac{6}{35}$ **6.** $5\frac{11}{22}$

$-\ 39\frac{7}{10}$ $-\ 24\frac{4}{5}$ $-\ 2\frac{7}{12}$

4. _____

5. _____

6. _____

7. $49\frac{1}{2}$ **8.** $24\frac{7}{27}$ **9.** $9\frac{1}{2}$

$-\ 29\frac{19}{36}$ $-\ 10\frac{5}{6}$ $-\ 7\frac{17}{18}$

7. _____

8. _____

9. _____

10. $8\frac{1}{2}$ **11.** $68\frac{5}{8}$ **12.** $44\frac{3}{16}$

$-\ 6\frac{5}{8}$ $-\ 9\frac{15}{16}$ $-\ 14\frac{19}{48}$

10. _____

11. _____

12. _____

13. _____

13. $29\frac{2}{3}$ **14.** $68\frac{5}{8}$ **15.** $88\frac{1}{3}$

$-\ 13\frac{1}{4}$ $-\ 36.37\frac{1}{2}$ $-\ 22\frac{1}{4}$

14. _____

15. _____

16. a. _____

16. a. How much must be added to $\frac{1}{4}$ to get $\frac{1}{2}$?

       b. _____

17. _____

18. _____

19. _____

b. How much must be added to $\frac{3}{8}$ to get $\frac{7}{8}$?

20. _____

Calculator Tip

To subtract mixed numbers using a calculator, write each number as a decimal by writing the fraction as a decimal and then adding the whole number.

17. $222\frac{8}{15}$ **18.** $10\frac{2}{3}$

$-\ 173\frac{21}{25}$ $-\ 5\frac{5}{6}$

19. $25\frac{17}{24}$ **20.** $26\frac{7}{20}$

$-\ 10\frac{31}{36}$ $-\ 24\frac{11}{14}$

Problems

21. $65\frac{8}{35}$

    $-\,16\frac{11}{14}$

22. $267\frac{4}{7}$

    $-\,203\frac{2}{9}$

21. _____

22. _____

23. _____

23. 82

    $-\,9\frac{5}{22}$

24. $173\frac{21}{25}$

    $-\,94\frac{8}{15}$

24. _____

25. _____

26. _____

25. $\frac{9}{19}$

    $-\,\frac{3}{17}$

26. $365\frac{27}{50}$

    $-\,123\frac{39}{40}$

27. _____

28. _____

27. 47

    $-\,8\frac{3}{27}$

28. $\frac{11}{15}$

    $-\,\frac{7}{10}$

29. Dr. White has a patient who needs to reach his desired weight of 182 lb. To reach this weight, the patient needs to lose 15 lb. If the patient loses $5\frac{1}{2}$ lb the first week and $3\frac{1}{4}$ lb the second week, how much more weight must he lose to reach his desired weight?

29. _____

Business Applications

30. Hutchens Recycling Company collects steel to be recycled. To determine how much steel the company collected, subtract the amount in the Weight of Cart column from the gross weight to find the weight of the load in each cart. Then find the total weight of the five loads.

	Cart No.	Gross Weight	Weight of Cart	Weight of Load
a.	15	$3,856\frac{1}{2}$ lb	$1,609\frac{7}{8}$ lb	_____
b.	22	$4,516$ lb	$1,637\frac{3}{4}$ lb	_____
c.	26	$4,763\frac{1}{8}$ lb	$1,609\frac{7}{8}$ lb	_____
d.	31	$4,873\frac{1}{4}$ lb	$1,609\frac{7}{8}$ lb	_____
e.	35	$4,529\frac{7}{8}$ lb	$1,637\frac{3}{4}$ lb	_____
			Total weight	_____

Answer

31. Gem Industries' stock closed on Friday at a price of $42\frac{5}{8}$. On Monday, this stock cosed at a price of $39\frac{7}{8}$. By how much did the market value of the stock drop on Monday?

31. _____

32. Raymond's Fish Hatchery made a shipment to the locker plant. To determine the weight of fish sent, subtract the bucket weight from the gross weight of each catch to find the weight of the fish. Then find the total of the amounts in the Weight of Fish column.

	Gross Weight	Bucket Weight	Weight of Fish
a.	$327\frac{5}{16}$ lb	$42\frac{1}{2}$ lb	_____
b.	$321\frac{5}{8}$ lb	$42\frac{1}{2}$ lb	_____
c.	$337\frac{7}{8}$ lb	$45\frac{3}{4}$ lb	_____
d.	$344\frac{1}{4}$ lb	$45\frac{3}{4}$ lb	_____
e.	$323\frac{3}{8}$ lb	$42\frac{1}{2}$ lb	_____
		Total weight	_____

33. Convert the men's race times to seconds, changing the decimals to fractions. Compute the difference in the times. Do you agree with the 17.2 second difference quoted below?

33. _____

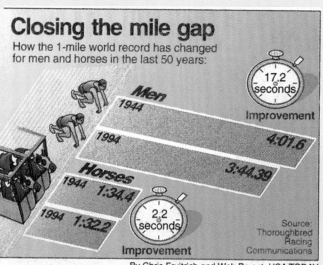

Closing the mile gap

How the 1-mile world record has changed for men and horses in the last 50 years:

Men 1944
1994
17.2 seconds Improvement
4:01.6
3:44.39

Horses
1944 1:34.4
1994 1:32.2
2.2 seconds Improvement

Source: Thoroughbred Racing Communications

By Chris Fruitrich and Web Bryant, USA TODAY

SKILLBUILDER 4.5

Multiplying Fractions and Mixed Numbers

Learning Outcomes

After completing this skillbuilder, you should be able to:

- Find the product of two or more fractions.
- Find the product of two or more mixed numbers.

Multiplying Fractions

To multiply fractions, there is no need to convert denominators to like terms; any combination of factors can be multiplied. Multiply numerator by numerator and denominator by denominator. Rename the resulting product in simplest terms. If an improper fraction results, change it to a mixed or whole number.

If **cross factors** (the numerator of one fraction and the denominator of the other fraction) can be evenly divided before multiplying, do so in order to simplify the multiplication. This process of dividing cross factors is called **cancellation.** If either the multiplicand or the multiplier is a whole number, express the whole number with a denominator of 1.

Problems

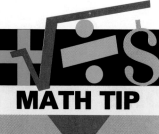

MATH TIP

Here are two more divisibility rules you can use in simplifying fractions before multiplying: A number is divisible by 4 if the last two digits in the number form a number divisible by 4 and by it's divisible by 8 if the last three digits form a number divisible by 8.

EXAMPLE

a. Multiply: $\frac{3}{8} \times \frac{2}{9}$.

b. Multiply: $5 \times \frac{2}{15}$.

SOLUTION

a. $\dfrac{\overset{1}{\cancel{3}}}{\underset{4}{\cancel{8}}} \times \dfrac{\overset{1}{\cancel{2}}}{\underset{3}{\cancel{9}}} = \dfrac{1}{12}$

b. $5 \times \dfrac{2}{15} = \dfrac{\overset{1}{\cancel{5}}}{1} \times \dfrac{2}{\underset{3}{\cancel{15}}} = \dfrac{2}{3}$

▶ Self-Check

a. Multiply: $\frac{3}{5} \times \frac{5}{8}$.

b. Multiply: $12 \times \frac{2}{9}$.

Self-Check Answers

a. ____$\frac{3}{8}$____

b. ____$2\frac{2}{3}$____

Find the products in the following problems. Rename the answers in simplest terms.

Answers

1. $\dfrac{5}{16} \times \dfrac{4}{15} =$ **2.** $\dfrac{8}{9} \times \dfrac{5}{24} =$ **3.** $\dfrac{7}{12} \times \dfrac{8}{49} =$

1. _____

2. _____

3. _____

4. $\dfrac{5}{6} \times \dfrac{3}{4} =$ **5.** $\dfrac{4}{7} \times \dfrac{3}{7} =$ **6.** $3 \times \dfrac{5}{12} =$

4. _____

5. _____

6. _____

7. $\dfrac{8}{27} \times 9 =$ **8.** $6 \times \dfrac{5}{18} =$ **9.** $\dfrac{3}{5} \times \dfrac{2}{3}$

7. _____

8. _____

9. _____

Problems

10. $\dfrac{2}{5} \times \dfrac{3}{10} =$

11. $5 \times \dfrac{3}{25} =$

12. $\dfrac{5}{8} \times \dfrac{9}{16}$

13. $\dfrac{1}{2} \times \dfrac{3}{8} =$

14. $7 \times \dfrac{4}{21} = \dfrac{4}{3} =$

15. $\dfrac{11}{25} \times \dfrac{10}{33} =$

16. $\dfrac{8}{21} \times \dfrac{3}{16} \times \dfrac{2}{5} =$

17. $\dfrac{5}{34} \times \dfrac{4}{15} \times \dfrac{7}{8} =$

18. $\dfrac{8}{51} \times \dfrac{3}{16} \times \dfrac{5}{12} =$

19. Derek, Andrea, and Keil formed a partnership to market Worldwide Distributors. They agreed to take a share of the profits each month for personal expenses and to put the balance back into the business. The shares agreed upon were $\dfrac{2}{5}$, $\dfrac{1}{4}$, and $\dfrac{1}{8}$, respectively. The gross profit for the month of June was $22,520.

a. What fraction of the gross profit did their combined shares equal?

b. What fraction of the gross profit did they put back into the business?

c. What was the value of each of their shares?

10. _____

11. _____

12. _____

13. _____

14. _____

15. _____

16. _____

17. _____

18. _____

19. a. _____

b. _____

c. _____

_____

_____

Problems

20. Rabin Market's monthly expenses are $24,000. Of this $24,000, $\frac{1}{3}$ is for salaries, $\frac{1}{16}$ is for utilities, $\frac{1}{8}$ is for advertising, and $\frac{1}{12}$ is for general maintenance. What is the expense each month for the following?

a. Salaries
b. Utilities
c. Advertising
d. General maintenance

21. Hilary Antell operates a chicken farm on part of her 624-acre property. The house, barn, and roads presently occupy $\frac{1}{5}$ of the total acreage; chicken coops and runs occupy $\frac{2}{9}$ of the total. She plans to devote $\frac{1}{4}$ of the total acreage to fruit trees and the balance to vegetable fields.

a. How many acres will the fruit trees occupy?
b. What fraction of the total will be devoted to vegetable fields?

22. Andrew Porter died without leaving a will. Under the laws of his state, $\frac{1}{2}$ of his personal estate goes to his widow and the remainder is divided equally among his eight children. The estate has a value of $284,800.

a. How much would Porter's widow receive?
b. How much would each child receive?

Calculator Tip

Memory keys allow you to store amounts for later use.

Problems

23. The City of Laird has an opportunity to purchase land for a park. The section in question measures $\frac{7}{8}$ mile by $\frac{5}{12}$ mile.

a. If there are 640 acres to the square mile, how many acres are contained in the plot?

b. When federal funds were withdrawn, the city had only enough money to buy $77\frac{1}{9}$ acres of the available land. How many acres were they unable to buy?

Answers

23. a.————————

   b.————————

24. a.————————

   b.————————

24. A rule of thumb says that your monthly mortgage payment, including interest, should not exceed $\frac{7}{25}$ of your gross monthly income.

a. Bruce Angler earns $43,500 a year. How much of a mortgage payment can he afford per month?

b. Yvette Saunders earns $51,800 a year. How much of a mortgage payment can she afford per month?

Multiplying Mixed Numbers

To multiply mixed numbers, rename the factors as improper fractions. Use cancellation if possible, and rename the answer in lowest terms.

Decimal fractions may be multiplied by common fractions and mixed numbers.

Problems

Find the product in each of the following problems. Show the details of your work, and check each product.

Answers

25. $12\frac{1}{4} \times 2\frac{2}{7} =$　　　　　　**26.** $24\frac{1}{4} \times 15\frac{1}{3} =$

27. $36\frac{2}{3} \times 21\frac{3}{8} =$　　　　　　**28.** $12\frac{7}{8} \times 6\frac{3}{4} =$

29. $3\frac{1}{3} \times 4\frac{1}{8} =$　　　　　　**30.** $6\frac{3}{4} \times 1\frac{1}{9} =$

25. _____

26. _____

27. _____

28. _____

29. _____

30. _____

Problems

31. $8\frac{1}{6} \times 2\frac{1}{7} =$ **32.** $11\frac{1}{4} \times 1\frac{1}{15} =$

Answers

31. _____

32. _____

33. _____

34. _____

33. $1\frac{1}{2} \times 2\frac{5}{6} =$ **34.** $6\frac{1}{2} \times 3\frac{1}{4} =$

35. _____

36. _____

35. $2\frac{1}{3} \times 3\frac{1}{2} =$ **36.** $3\frac{1}{3} \times 4\frac{1}{5} =$

37. Last year Carlos Larson jogged $3\frac{5}{8}$ mi each day. Now he   **37.** _____
jogs $2\frac{1}{2}$ times this distance. How many miles does he now run
each day?

38. Harry Wilson uses his car for business. Last year he drove   **38.** _____
21,876 mi, using $1,393\frac{4}{10}$ gal of gasoline. If the average price per gallon was $\$1.35\frac{9}{10}$ a gallon, how much did he
spend on gasoline? (Round your final answer to the nearest
cent.)

Problems

39. The Hardings are ordering new carpeting costing $24.75 a square yard ($2.75 a square foot). The areas to be carpeted are the living room, $21\frac{1}{2}$ ft long by $18\frac{1}{4}$ ft wide; and the dining room, $18\frac{1}{4}$ ft long by $14\frac{3}{4}$ foot wide. What will be the total cost of carpet for these rooms? (Multiply length times width to get area. Determine the total area first; then work out the cost. Round your answer to the nearest cent.)

39. _____

40. For each pound of pistachio nuts purchased, Taft's Nuts is offering customers an extra $\frac{1}{4}$ lb free. Alice Tam bought 5 lb of the nuts. How many pounds will she get free?

40. _____

41. Lucy Ford has been hired by the Carter Corporation to do odd jobs. She will be paid $8.50 an hour. On Monday, she worked $4\frac{1}{2}$ hr. On Tuesday, she worked $6\frac{1}{4}$ hr. On Friday, she worked $3\frac{3}{4}$ hr. How much did she earn for the 3 days that she worked?

41. _____

Business Applications

42. Widget Industries has rented booth space at the International Trade Show Center. In the past, Widget has spent $\frac{7}{16}$ of every dollar on exhibit construction and space rental, $\frac{1}{8}$ for transportation, $\frac{11}{32}$ on specialty advertising and personnel, and the balance on cleanup and removal of the exhibit after the show. If Widget has budgeted $228,000 for the coming show, how much will it spend on each of these items?

42. _____

_____

_____

_____

43. The current market values of May Garcia's stocks are listed here. What is the value of her investment in each stock, and what is the total value of her portfolio? (In computing your answers, change any fractional parts of a dollar to cents, rounding off to three decimal places, where necessary.)

a. Stock A   54 shares at 48\frac{1}{8}$ a share

b. Stock B   300 shares at 12\frac{5}{8}$ a share

c. Stock C   80 shares at 39\frac{1}{2}$ a share

d. Stock D   84 shares at 68\frac{7}{8}$ a share

e. Stock E   312 shares at 32\frac{3}{4}$ a share

f. Total

Answers

43. **a.** _____

b. _____

c. _____

d. _____

e. _____

f. _____

Business Applications

44. Figure the gross pay on the Merry Maids' payroll register.

	Employee	Hours Worked	Hourly Rate	Gross Pay
a.	Brittain	$38\frac{1}{4}$	$5.88	
b.	Jones	$37\frac{1}{2}$	$6.75	
c.	Ming	$38\frac{5}{6}$	$6.55	
d.	Cummings	$39\frac{1}{4}$	$6.87\frac{1}{2}$	
e.	Valdez	$38\frac{3}{4}$	$7.75	
f.	Huff	39	$5.37\frac{1}{2}$	
g.	Routh	$37\frac{1}{3}$	$7.87\frac{1}{2}$	

Student Notes

SKILLBUILDER 4.6

Dividing Fractions and Mixed Numbers

Learning Outcomes

After completing this skillbuilder, you should be able to:

- Divide one fraction by another.
- Divide one mixed number by another.

Dividing Fractions

To divide common fractions, invert (turn upside down) the divisor and multiply. Rename the quotient in simplest terms. When either the dividend or the divisor is a whole number, write it as that number over 1 before dividing. Do not use cancellation until *after* the divisor has been inverted.

MATH TIP

A number is divisible by 6 if the number is divisible by both 2 and 3; by 9 if the sum of the digits is divisible by 9; and by 10 if the number ends in a 0.

EXAMPLE

Find each quotient.

a. $\dfrac{3}{5} \div \dfrac{9}{16}$

b. $\dfrac{5}{12} \div 3$

SOLUTION

a. $\dfrac{3}{5} \div \dfrac{9}{16} = \dfrac{\cancel{3}}{5} \times \dfrac{16}{\cancel{9}_{3}} = \dfrac{16}{15} = 1\dfrac{1}{15}$

b. $\dfrac{5}{12} \div 3 = \dfrac{5}{12} \div \dfrac{3}{1} = \dfrac{5}{12} \times \dfrac{1}{3} = \dfrac{5}{36}$

▶ **Self-Check**

Find the quotient: $4 \div \dfrac{2}{9}$.

Self-Check Answer

_____18_____

Problems

Find the quotients in the following problems. Rename the answers in simplest terms if necessary.

1. $\dfrac{3}{8} \div \dfrac{7}{9}$

2. $\dfrac{2}{7} \div \dfrac{4}{11}$

3. $\dfrac{5}{12} \div 15$

4. $\dfrac{5}{6} \div \dfrac{5}{12}$

5. $\dfrac{3}{16} \div \dfrac{5}{8}$

6. $25 \div \dfrac{5}{9}$

7. $\dfrac{16}{35} \div \dfrac{2}{7}$

8. $\dfrac{5}{24} \div \dfrac{3}{10}$

9. $12 \div \dfrac{3}{8}$

10. $\dfrac{14}{17} \div 21$

11. $\dfrac{3}{4} \div \dfrac{9}{32}$

12. $\dfrac{12}{57} \div \dfrac{2}{3}$

13. $\dfrac{5}{9} \div \dfrac{5}{18}$

14. $\dfrac{2}{5} \div \dfrac{28}{31}$

15. $\dfrac{5}{36} \div \dfrac{4}{9}$

16. $\dfrac{9}{16} \div \dfrac{7}{18}$

17. $42 \div \dfrac{18}{25}$

18. $\dfrac{7}{8} \div \dfrac{5}{32}$

1. _____
2. _____
3. _____
4. _____
5. _____
6. _____
7. _____
8. _____
9. _____
10. _____
11. _____
12. _____
13. _____
14. _____
15. _____
16. _____
17. _____
18. _____

Problems

19. Environetics' design engineer has developed an energy-saving light requiring a platinum ribbon. The platinum ribbon comes from a roll 174 in. long. Each ribbon must be $\frac{3}{32}$ in. long. How many ribbons can be cut from the roll?

19. ——————

20. Rita Garland wishes to tile her 18-ft by 15-ft office with a 9-in. by 9-in. tile. How many whole tiles will she need to do the job?

20. ——————

21. Carlson Nursery has a garden plot measuring 162 ft by 162 ft. They plan to plant rows of seeds $\frac{3}{4}$ of a yard apart, planting as many rows as possible. How many rows will they be able to plant?

21. ——————

22. Maribel Donner owns Donner's Florals. She is cutting ribbon from 15-ft rolls. Each ribbon will be $\frac{3}{4}$ ft long. If she needs 220 ribbons in all, how many 15-ft rolls will she need?

22. ——————

23. A farmer sold 48 baskets of tomatoes for $4.50 a basket. The farmer used the proceeds, which amounted to $\frac{2}{3}$ of the total cost, toward the purchase of seedlings. How much did the farmer spend on seedlings?

23. ——————

Problems

Multiplying Mixed Numbers

To divide mixed numbers, first write them as improper fractions. Then invert the divisor and multiply. Before multiplying, be sure to perform any cancellation possible. Rename the resulting quotient in its simplest terms.

Decimal fractions may be divided by common fractions and mixed numbers.

EXAMPLE

a. Find the quotient: $15\frac{3}{8} \div 5\frac{1}{4}$.

b. Find the quotient: $\$15.85 \div 3\frac{4}{9}$.

SOLUTION

a. $15\frac{3}{8} \div 5\frac{1}{4} = \frac{123}{8} \div \frac{21}{4} = \frac{\overset{41}{\cancel{123}}}{\underset{2}{\cancel{8}}} \times \frac{\overset{1}{\cancel{4}}}{\underset{7}{\cancel{21}}} = \frac{41}{14} = 2\frac{13}{14}$

b. $\$15.85 \div 3\frac{4}{9} = \frac{\$15.85}{1} \div \frac{31}{9}$

$= \frac{\$15.85}{1} \times \frac{9}{31} = \frac{\$142.65}{31} = \$4.601 \rightarrow \$4.60$

▶ **Self-Check**

Find the quotient: $7\frac{2}{3} \div 4$.

Self-Check Answer

$1\frac{11}{12}$

Find the quotients in the following problems. Show the details of your work. In problems 36–41, you may want to change the decimal fractions to common fractions. If not, round your final answers to four decimal places, where necessary.

Answers

24. $32\frac{1}{6} \div 10\frac{2}{3}$ **25.** $24\frac{3}{5} \div 8\frac{1}{5}$

24. _____

25. _____

26. _____

27. _____

26. $5\frac{1}{8} \div 2\frac{2}{9}$ **27.** $1\frac{2}{3} \div 1\frac{2}{3}$

Problems

28. $16\frac{3}{4} \div 15\frac{2}{5}$ **29.** $12\frac{1}{2} \div 8\frac{1}{3}$

30. $12\frac{3}{5} \div \frac{3}{4}$ **31.** $36\frac{2}{3} \div 9$

32. $24\frac{1}{2} \div 8\frac{3}{4}$ **33.** $51\frac{5}{12} \div 12\frac{2}{5}$

34. $26 \div 3\frac{3}{8}$ **35.** $48\frac{1}{2} \div \frac{7}{8}$

36. $0.625 \div 1\frac{4}{9}$ **37.** $0.375 \div 5\frac{1}{2}$

38. $0.25 \div 22\frac{3}{8}$ **39.** $0.75 \div 41\frac{1}{3}$

Answers

28. _____
29. _____
30. _____
31. _____
32. _____
33. _____
34. _____
35. _____
36. _____
37. _____
38. _____
39. _____

Problems

40. $0.125 \div \frac{1}{3}$

41. $0.4 \div 12\frac{1}{4}$

Answers

40. _____

41. _____

42. Canteen Cooperative serves soup as part of its daily menu. The soup is packed in cans containing 46 oz each; the average portion size is $7\frac{1}{3}$ oz. On an average day, 125 portions of soup are served. To the nearest whole can, how many cans of soup are used each day?

42. _____

43. Richard Rodino paid $9,481.75 for the purchase of common stock. The purchase price was $48\frac{7}{8}$ a share. How many shares of stock did he purchase?

43. _____

44. One package of natural cereal costs 63¢ for $7\frac{1}{2}$ oz. Another costs $1.39 for $18\frac{3}{4}$ oz.

a. What is the price of one ounce to the nearest tenth of a cent in the smaller package?

b. What is the price of one ounce to the nearest tenth of a cent in the larger package?

44. a. _____

b. _____

© by Glencoe/McGraw-Hill.

Business Applications

45. Lloyd King is a top sales representative for Global Safety Products. Last month his sales expenses amounted to $4,800, which represented $\frac{3}{32}$ of his sales for the month.

45. a. _____

b. _____

a. What was the total of Lloyd's sales for the month?

b. Lloyd's commission is $\frac{2}{15}$ of his net sales (total sales less expenses) for the month. What was the amount of his commission, to the nearest cent, for that same month?

46. Betty Giddens works in the cutting rooms of an exclusive women's clothing manufacturer. Her next work order is a piece of fabric $128\frac{3}{8}$ yd long, from which she must cut skirts requiring $2\frac{1}{4}$ yd each. How many skirts will she be able to cut from this piece of fabric? (*Hint:* Round your answer down to the next lowest whole unit.)

46. _____

47. The city of Alexandria is planning a $22\frac{1}{4}$-mile long bus route. If a bus stop is placed every $1\frac{1}{2}$ mi, how many stops will be on the route? (Round your answer to the nearest whole number.)

47. _____

Business Applications

48. What was the average increase in working women per year? (Compute using mixed numbers.)

Answers

48. _____

Calculator Tip

The constant selector key lets you enter a constant divisor or factor only once and then use it repeatedly.

More women are working

If history holds true, most of the daughters who take part in "Take our Daughters to Work Day" will one day work themselves. Percentage of women who work:

57.8%

60%

40%

20%

0

'92

27.4%

WOMEN WORKING

'40

Source: Statistical Abstracts of the United States

By Cindy Hall and Marty Baumann, USA TODAY

Student Notes

SKILLBUILDER 4.7
Computing Fractional Parts

Learning Outcomes

After completing this skillbuilder, you should be able to:

- Use fractional parts to solve problems.
- Use fractional parts to find the number of items when given unit price and total cost.

Fractional Parts

Any number that can be divided evenly into another number is known as a **fractional part** of that number. Many computations can be simplified by recognizing the fractional parts of one or more of the terms and by using those values in place of the original figures in the computation.

For example, suppose you want to determine how much it would cost to purchase 1,264 metric converter slides at 25¢ each. Rather than multiply 1,264 by 25¢, you should recognize that 25¢ equals $\frac{1}{4}$ of $1. If the items sold for $1 each, the cost would be $1,264. However, since they sold for 25¢, or $\frac{1}{4}$ of a dollar, you can multiply $1,264 by $\frac{1}{4}$ to obtain a purchase cost of $316.

MATH TIP

Fractional parts make it possible to do many calculations mentally.

EXAMPLE

a. What part of $1, or 100¢ is $37\frac{1}{2}$¢?

b. Find the cost of 432 items at $37\frac{1}{2}$¢ each.

SOLUTION

a. Place $37\frac{1}{2}$¢ over $1.00 (or 100 cents) and rename the fraction in its simplest terms.

$$37\frac{1}{2} \div 100 = \frac{37\frac{1}{2}}{100} = \frac{2 \times 37\frac{1}{2}}{2 \times 100} = \frac{75}{200} = \frac{3}{8}$$

—or—

$$37\frac{1}{2} \div 100 = \frac{75}{2} \div \frac{100}{1} = \frac{\overset{3}{\cancel{75}}}{2} \times \frac{1}{\underset{4}{\cancel{100}}} = \frac{3}{8}$$

b. $37\frac{1}{2}$¢ $= \frac{3}{8}$ of $1

$$\frac{\overset{54}{\cancel{432}}}{1} \times \frac{3}{\underset{1}{\cancel{8}}} = \$162$$

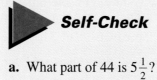 **Self-Check**

a. What part of 44 is $5\frac{1}{2}$?

b. Find the cost of 1,256 items at $37\frac{1}{2}$¢ each.

Self-Check Answer

$\dfrac{1}{8}$

$471

Problems

Find the fractional parts of $1 represented by the following amounts. Show details of your work.

1. 50¢ = _____ of $1   **2.** $6\frac{1}{4}$¢ = _____ of $1

3. $8\frac{1}{3}$¢ = _____ of $1   **4.** $33\frac{1}{3}$¢ = _____ of $1

5. $87\frac{1}{2}$¢ = _____ of $1   **6.** $37\frac{1}{2}$¢ = _____ of $1

7. 25¢ = _____ of $1   **8.** $12\frac{1}{2}$¢ = _____ of $1

9. 75¢ = _____ of $1   **10.** $16\frac{2}{3}$¢ = _____ of $1

11. $83\frac{1}{3}$¢ = _____ of $1   **12.** $62\frac{1}{2}$¢ = _____ of $1

Problems

Using the methods shown in this skillbuilder, find the cost of each of the following purchases.

Answers

13. 408 qt @ $16\frac{2}{3}$¢   **14.** 368 in @ $6\frac{1}{4}$¢   **15.** 735 in @ $33\frac{1}{3}$¢

13. _____

14. _____

15. _____

16. _____

17. _____

16. 252 in @ $12\frac{1}{2}$¢   **17.** 1,272 qt @ $66\frac{2}{3}$¢   **18.** 492 qt @ $8\frac{1}{3}$¢

18. _____

Finding the Number of Items

Fractional parts can be used to find the quantity of items bought when the unit price and the total cost are known.

Calculator Tip

Memorize the most common fractional parts of $1. This simplifies calculations on the calculator.

EXAMPLE

Jean McCrea, a candidate for public office, spent $108 for monogrammed pencils to be given to prospective voters. The pencils cost $12\frac{1}{2}$¢ each. How many pencils were bought?

SOLUTION

$$12\frac{1}{2}¢ = \frac{1}{8} \text{ of } \$1 \qquad \$108 \div \frac{1}{8} = \frac{108}{1} \times \frac{8}{1} = 864 \text{ pencils}$$

 **Self-Check**

Cotton cloth costs $87\frac{1}{2}$¢ a yard. If a total purchase came to $203, how many yards were bought?

Self-Check Answer

_____ 232 yd _____

NAME: _____ DATE: _____

Problems

19. $70 for decals at $8\frac{1}{3}$ ¢ each

19. _____

20. _____

20. $213.50 for paint thinner at $87\frac{1}{2}$ ¢ a pint

21. _____

22. _____

23. _____

21. $135 for felt at $93\frac{3}{4}$ ¢ a meter

22. $168 for rice flour at 50¢ a pound

23. $225 for orange juice at 75¢ a quart

Business Applications

24. Serena Rolfe, owner of Rolfe's Auto Repair, purchased automatic transmission fluid for her shop storeroom. Ms. Rolfe paid $1.25 a quart for automatic transmission fluid. Her total bill came to $15.00. How many quarts did she purchase?

Answers

24. _____

25. Tariq Keller is in charge of purchasing supplies for High Performance Auto Supply Store. When reordering brake fluid, Mr. Keller paid 87.5¢ a pint. The total bill came to $71.75. How many pints did he purchase?

25. _____

26. During monthly inventory at Routh's Pharmacy, Ms. Routh decided she needed to restock her shelves with vitamin C tablets. She paid $90.00 for a case of tablets. If each bottle cost $3.75, how many bottles did she purchase?

Answers

26. _____

27. If Sanders bats against left-handers 15 times, how many hits should he expect?

27. _____

Lefties challenge Sanders

Career batting average of Atlanta's left-hand-hitting Deion Sanders, playing full-time this season:

Against left-handers .191

Against right-handers .266

Source: STATS 1994 Players Profiles

By John Riley and Marcia Staimer, USA TODAY

Student Notes

_____

_____

_____

_____

_____

_____

_____

_____

_____

_____

_____

SKILLBUILDER **4.8**

Using Fractional Parts

Learning Outcomes

After completing this skillbuilder, you should be able to:

■ Use fractional parts to speed up computations.

Using Fractional Parts

Fractional parts are useful only if their values and relationships are recognized quickly. The following list shows the fractional parts of $1 that are used most frequently. Use it as necessary for the problems.

$$\$\frac{1}{2} = 50¢ \qquad \$\frac{1}{5} = 20¢ \qquad \$\frac{1}{8} = 12\frac{1}{2}¢ \qquad \$\frac{7}{10} = 70¢$$

$$\$\frac{1}{3} = 33\frac{1}{3}¢ \qquad \$\frac{2}{5} = 40¢ \qquad \$\frac{3}{8} = 37\frac{1}{2}¢ \qquad \$\frac{9}{10} = 90¢$$

$$\$\frac{2}{3} = 66\frac{2}{3}¢ \qquad \$\frac{3}{5} = 60¢ \qquad \$\frac{5}{8} = 62\frac{1}{2}¢ \qquad \$\frac{1}{12} = 8\frac{1}{3}¢$$

$$\$\frac{1}{4} = 25¢ \qquad \$\frac{4}{5} = 80¢ \qquad \$\frac{7}{8} = 87\frac{1}{2}¢ \qquad \$\frac{1}{16} = 6\frac{1}{4}¢$$

$$\$\frac{3}{4} = 75¢ \qquad \$\frac{1}{6} = 16\frac{2}{3}¢ \qquad \$\frac{1}{10} = 10¢ \qquad \$\frac{1}{20} = 5¢$$

$$\$\frac{5}{6} = 83\frac{1}{3}¢ \qquad \$\frac{3}{10} = 30¢$$

Fractional parts can be used to speed up computations for values greater than $1 too.

Problems

MATH TIP

Memorizing the fractional parts you use most frequently will save you time when calculating fractions.

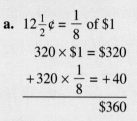

EXAMPLE

Find the total cost.

a. 320 pieces at 1.12\frac{1}{2}$ each.

b. 256 units at $37.50 each.

SOLUTION

a. $12\frac{1}{2}¢ = \frac{1}{8}$ of $1

$320 \times \$1 = \$320$

$+320 \times \dfrac{1}{8} = +40$

$\overline{\phantom{+320 \times \dfrac{1}{8} = }\$360}$

b. $37.50 = \frac{3}{8}$ of $100

$\dfrac{256}{1} \times \dfrac{3}{8} = 96$

$96 \times 100 = \$9,600$

 **Self-Check**

Find the cost of 112 items at $6.25 each.

Self-Check Answer

$700

Find the total cost of each of the following groups of purchases. In the parentheses after each price, show what fractional part of $1 that price represents. Round your answers to the nearest cent where necessary.

1. 128 qt @ $12\frac{1}{2}¢$ ( ) = _____

   360 qt @ $16\frac{2}{3}¢$ ( ) = _____

   252 qt @ $83\frac{1}{3}¢$ ( ) = _____

   468 qt @ 40¢ ( ) = _____

          Total _____

2.   355 ft @ 60¢ ( ) = _____

   288 ft @ $12\frac{1}{2}¢$ ( ) = _____

   369 ft @ $33\frac{1}{3}¢$ ( ) = _____

   127 ft @ 75¢ ( ) = _____

          Total _____

3. 300 yd @ $62\frac{1}{2}¢$ ( ) = _____

   364 yd @ 25¢ ( ) = _____

   112 yd @ $37\frac{1}{2}¢$ ( ) = _____

   415 yd @ $33\frac{1}{3}¢$ ( ) = _____

          Total _____

4. 286 lb @ $66\frac{2}{3}¢$ ( ) = _____

   280 lb @ 20¢ ( ) = _____

   720 lb @ $66\frac{2}{3}¢$ ( ) = _____

   752 lb @ $6\frac{1}{4}¢$ ( ) = _____

          Total _____

Problems

Calculator Tip

A conversion chart is often helpful when using a calculator to solve problems involving fractional parts.

Find the cost in each of the following problems.

a. 320 items @ $37\frac{1}{2}$¢ (   )

b. 560 items @ $62\frac{1}{2}$¢ (   )

c. 200 items @ 5¢ (   )

d. 550 items @ 80¢ (   )

e. 32 items @ $12\frac{1}{2}$¢ (   )

f. 48 items @ 25¢ (   )

Answers

5. **a.** _____

   b. _____

   c. _____

   d. _____

   e. _____

   f. _____

Compute the extensions on the following purchases using fractional parts for all values. In the parentheses after the unit price, show the fraction of $10, $100, and so on, that the price represents. Round to the nearest cent where necessary. Then find the total.

6.   42 jackets at $62.50 each (   ) = _____

    74 cotton shirts at $12.50 each (   ) = _____

       116 pairs socks at $6.25 each (   ) = _____

         15 pairs jeans at $25.00 each (   ) = _____

           48 T-shirts at $8.33 each (   ) = _____

    13 sport jackets at $90.00 each (   ) = _____

                                    Total   _____

Business Applications

The accountant for Tapes, Etc., Arleen's Art Shop, and Mechanic's World completes charts like the following to show sales for the month of May. Compute the extensions using fractional parts to determine monthly sales for each item. Show the fraction of $10, $100, and so on, that the price represents. Then find the totals.

TAPES, ETC.

7.

18 radios at $37.50 each ( ) = _____

15 compact disc cases at $8.33 each ( ) = _____

12 tapes at $6.25 each ( ) = _____

23 compact discs at $8.75 each ( ) = _____

60 carrying cases at $8.33 each ( ) = _____

87 video tapes at $3.75 each ( ) = _____

Total _____

ARLEEN'S ART SHOP

8.

46 brushes at $6.25 each ( ) = _____

48 paint boxes at $9.00 each ( ) = _____

32 cans paint thinner at $6.25 each ( ) = _____

24 clay pots at $16.67 each ( ) = _____

18 table easels at $8.75 each ( ) = _____

20 palettes at $12.50 each ( ) = _____

10 sheets sand paper at $0.50 each ( ) = _____

Total = _____

Business Applications

MECHANIC'S WORLD

9. 26 6-piece wrench sets at $37.50 each (   ) = _____

52 drills at $12.50 each (   ) = _____

318 screw packs at $6.25 each (   ) = _____

6 small socket sets at $37.50 each (   ) = _____

12 entrance locks at $16.67 each (   ) = _____

15 latch sets at $8.33 each (   ) = _____

10 channel locks at $6.25 each (   ) = _____

Total = _____

Student Notes

_____

_____

_____

_____

_____

_____

_____

_____

_____

_____

_____

_____

_____

_____

_____

U N I T

5

PERCENTS

Percents are probably one of the most common ways to show data because they give us a means of determining what part of the whole is involved. They also allow us to compare data from different-sized groups. This chart from *The Wall Street Journal* shows the results of a survey of how hourly and salaried workers view their jobs.

In this unit we study percents and their application to commissions and to markup and markdown. In addition, we study graphs, tables, ratios, and proportions.

HOURLY VS. SALARIED WORKERS: HOW THEY VIEW THEIR JOBS

A survey of hourly and salaried workers revealed the following:

- 52% of hourly workers didn't see a clear link between good performance and pay increases vs. 38% of salaried workers.
- 45% of hourly workers were satisfied with their pay vs. 53% of salaried workers.
- 54% of hourly workers were satisfied with their benefits vs. 64% of salaried workers.
- 35% of hourly workers felt secure in their jobs vs. 46% of salaried workers.
- 33% of hourly workers saw their company as more than "just a place to work" vs. 46% of salaried workers.
- 47% of hourly workers would recommend their company as a place to work vs. 58% of salaried workers.
- 54% of hourly workers rated their work environments positively vs. 65% of salaried workers.

Source: Wyatt Co.

Crew Manager

Joe Spencer is a crew manager for Capital Builders, a residential construction firm. Joe is responsible for ordering supplies, scheduling work crews, and inspecting the workmanship to ensure that it meets federal, state, and local standards. When a problem arises, Joe must decide on a solution. Joe spends most of his work days traveling from one job site to another, making sure that all of the homes are progressing according to schedule.

Math Application

Joe submits daily time sheets to his central office. The time sheet lists each job site, the time Joe spent at the site, and Joe's total hours for the day. Following is Joe's time sheet for Monday, February 7.

Job Site	Hours at Site
Long St.	$1\frac{1}{2}$
Main St.	$2\frac{1}{4}$
Franklin Ave.	$3\frac{1}{2}$

Total hours on site

Total hours, including travel time ($2\frac{3}{4}$ hrs)

What percent of the day did Joe spend at each job? Divide the time spent at each job site by the total hours on site.

Long St. $1\frac{1}{2} \div 7\frac{1}{4} = \underline{\hspace{1.5cm}}$

Main St. $2\frac{1}{4} \div 7\frac{1}{4} = \underline{\hspace{1.5cm}}$

Franklin Ave. $3\frac{1}{2} \div 7\frac{1}{4} = \underline{\hspace{1.5cm}}$

Critical Thinking Problem

Joe Spencer is scheduling work crews for the next two weeks. How might Joe project his total costs for the owners?

SKILLBUILDER 5.1

Percents and the Percentage Formula

Learning Outcomes

After completing this skillbuilder, you should be able to:

- Rename percents as decimals and fractions.
- Rename decimals and fractions or mixed numbers as percents.
- Use the percentage formula to find the percentage.

Renaming Percents as Decimals and Fractions

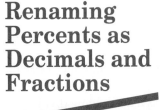

MATH TIP

The number of decimal places indicates the power of 10 in the denominator when you rename a percent as a fraction. Always reduce the fraction to lowest terms.

Percent means *per hundred*. Percents can be expressed as decimals or as common fractions or mixed numbers. To rename a percent as a decimal, divide the percent by 100 (or move the decimal point two places to the left) and remove the percent sign.

To rename a percent as a common fraction, first express the percent as a decimal fraction. Then rename the decimal as a common fraction reduced to lowest terms.

EXAMPLE

a. Rename 57% as a decimal.
b. Rename $8\frac{3}{4}\%$ as a common fraction reduced to lowest terms

SOLUTION

a. $57\% = 0\underset{\smile}{57}. = 0.57$

b. $8\frac{3}{4}\% = 0.0875 = \dfrac{875}{10,000} = \dfrac{7}{80}$

 **Self-Check**

a. Rename 142% as a decimal.
b. Rename 150% as a common fraction or mixed number in lowest terms.

Self-Check Answers

a. ____1.42____

b. ____$1\frac{1}{2}$____

Problems

Rename each percent as a decimal.

1. 24% **2.** 93% **3.** 17%

4. 8% **5.** 20.2% **6.** 158%

7. 0.5% **8.** 200% **9.** 1.15%

Rename each percent as a fraction or mixed number. Reduce to lowest terms, where necessary.

10. 25% = _____   **11.** 35% = _____   **12.** 16% = _____

13. 7% = _____   **14.** 45% = _____   **15.** $33\frac{1}{3}$% = _____

16. 15% = _____   **17.** $62\frac{1}{2}$% = _____   **18.** $6\frac{1}{4}$% = _____

19. 225% = _____   **20.** $83\frac{1}{3}$% = _____   **21.** $37\frac{1}{2}$% = _____

22. $2\frac{1}{4}$% = _____   **23.** $\frac{1}{4}$% = _____   **24.** 375% = _____

25. $87\frac{1}{2}$% = _____

Answers

1. _____

2. _____

3. _____

4. _____

5. _____

6. _____

7. _____

8. _____

9. _____

Renaming Decimals and Fractions as Percents

To rename a decimal as a percent, multiply the decimal by 100 and add a percent sign.

To rename a fraction as a percent, rename the fraction as a decimal by dividing the numerator by the denominator, and then rename the decimal as a percent.

EXAMPLE

a. Rename 0.333 as a percent.

b. Rename $\frac{1}{4}$ and $1\frac{1}{8}$ as percents.

SOLUTION

a. $0.333 = 0.333\% = 33.3\%$

b. $\frac{1}{4} = 0.25 = 0.25\% = 25\%$

$1\frac{1}{8} = 1.125 = 1.125\% = 112.5\%$

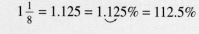 **Self-Check**

a. Rename 0.45 as a percent.

b. Rename $\frac{5}{8}$ as a percent

Self-Check Answers

a. ____45%____

b. ____$62\frac{1}{2}\%$____

Problems

Rename each common fraction as (a) a decimal fraction (rounded off to the nearest ten thousandth where necessary) and then to (b) a percent.

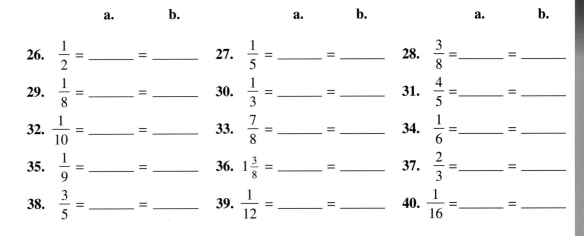

	a.	b.			a.	b.			a.	b.
26. $\frac{1}{2}$ =	_____	= _____	**27.** $\frac{1}{5}$ =	_____	= _____	**28.** $\frac{3}{8}$ =	_____	= _____
29. $\frac{1}{8}$ =	_____	= _____	**30.** $\frac{1}{3}$ =	_____	= _____	**31.** $\frac{4}{5}$ =	_____	= _____
32. $\frac{1}{10}$ =	_____	= _____	**33.** $\frac{7}{8}$ =	_____	= _____	**34.** $\frac{1}{6}$ =	_____	= _____
35. $\frac{1}{9}$ =	_____	= _____	**36.** $1\frac{3}{8}$ =	_____	= _____	**37.** $\frac{2}{3}$ =	_____	= _____
38. $\frac{3}{5}$ =	_____	= _____	**39.** $\frac{1}{12}$ =	_____	= _____	**40.** $\frac{1}{16}$ =	_____	= _____

Using The Percentage Formula

The **percentage formula** is

$$\text{Base} \times \text{Rate} = \text{Percentage}$$

A percent is also called a **rate.** When a rate is multiplied by a number, called the **base,** the result is a **percentage.**

Calculator Tip

If you misplace the decimal point, your percentage will be incorrect. Estimate your answer to check if it is logical.

EXAMPLE

What is $62\frac{1}{2}\%$ of 120?

SOLUTION

$62\frac{1}{2}\% = 0.625$

$\text{Base} \times \text{Rate} = \text{Percentage}$
$120 \times 0.625 = 75$

 **Self-Check**

What is 30% of $615?

Self-Check Answer

_____ $184.50

Problems

Compute the percentage in the following problems using the formula *base × rate = percentage* (*B × R = P*). Use the fractional equivalent for the rate (percent) whenever it will help solve the problem.

41. 8% of $500   **42.** 5% of $42.80   **43.** 25% of $5,800

44. 2% of 90   **45.** 72% of 613   **46.** $8\frac{1}{3}$% of 288

47. $12\frac{1}{2}$% of $8,010   **48.** $2\frac{1}{2}$% of $340   **49.** $16\frac{2}{3}$% of 264

50. $37\frac{1}{2}$% of 4,260   **51.** 9% of 1,854

52. $31\frac{1}{4}$% of $2,500   **53.** 52% of 18,500

54. $11\frac{1}{9}$% of 810   **55.** $7\frac{1}{2}$% of $7,800

56. $62\frac{1}{2}$% of 9,000   **57.** $41\frac{3}{4}$% of 864

58. $87\frac{1}{2}$% of 12,800   **59.** $43\frac{3}{4}$% of 795

60. $33\frac{1}{3}$% of $7,218   **61.** 21% of 6,320

41. _____
42. _____
43. _____
44. _____
45. _____
46. _____
47. _____
48. _____
49. _____
50. _____
51. _____
52. _____
53. _____
54. _____
55. _____
56. _____
57. _____
58. _____
59. _____
60. _____
61. _____

Business Applications

62. Last month, Roberta Jameson was paid a commission of 5% on sales totaling $78,000. How much did she earn for the month?

62. _____

63. Albert Graham received an $8\frac{1}{3}$% increase in his $25,848 annual salary.
  a. What was the total amount of his increase?
  b. How much did his monthly salary increase?
  c. What is his new annual salary?

63. **a.** _____

b. _____

c. _____

64. Emma Channing, sales manager for Queens' Cosmetics, is projecting an $11\frac{1}{9}$% increase in sales for the next quarter. Sales for this quarter were $412,893.
  a. What is the expected amount of increase for the next quarter's sale?
  b. If the increase in attained, what will total sales be for the next quarter?

64. **a.** _____

b. _____

Business Applications

65. José Martinez, marketing manager for Queen's Cosmetics, conducted a market survey and determined that 37.5% of the 2,150 people questioned recognized at least one product produced by Queen's Cosmetics. How many people surveyed knew about their products?

Answers

65. _____

66. If 1,500 men and women were surveyed, how many of each feel guilty?

66. _____

_____

Guilt for a day

Percentage of people who say they feel guilty about not buying Mother's Day or Father's Day gifts:

MEN 46.2%

WOMEN 33.6%

Source: Bruskin Goldring Research survey for Godiva Chocolatier

By Cindy Hall and Gary Visgaitis, USA TODAY

Student Notes

_____
_____
_____
_____
_____
_____
_____
_____
_____
_____
_____

SKILLBUILDER 5.2

Computing the Rate, Base, Increase, and Decrease

Learning Outcomes

After completing this skillbuilder, you should be able to:

- Use the percentage formula to compute base and rate.
- Compute either the amount or the rate of increase.
- Compute either the amount or the rate of decrease.

Using The Percentage Formula

The basic percentage formula can be used to find either the base or the rate. The following diagram is useful.

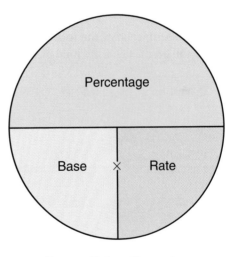

Base × Rate = Percentage
B × R = P

Cover the word representing the unknown element in the formula. The remaining portions of the chart indicate what operation is necessary to find the unknown. For example, to find 30% of 340, cover the word *percentage* since it is unknown. The diagram shows that you should multiply the base times the rate.

If the base is unknown, cover the word *base*. The diagram shows that you divide the percentage by the rate. If the rate is unknown, divide the percentage by the base.

Problems

MATH TIP

Remember the percentage is a part of the base (the whole) and the percent is the rate of that part; it always has a percent sign (%) after it.

EXAMPLE

a. 102 is 30% of what number?
b. What percent of 520 is 130?

SOLUTION

a. $P \div R = B$
$102 \div 0.30 = 340$
b. $130 \div 520 = 0.25 = 25\%$

Self-Check

a. 48% of ___?___ is 120.
b. What percent of 56 is 7?

Self-Check Answers

a. ____250____

b. ____$12\frac{1}{2}\%$____

Find the base in each of the following problems. Remember to rename the percent as a decimal or fraction.

Answers

1. 25% of $___?___ = $144

2. 30% of $___?___ = $21.90

3. 40% of $___?___ = $75.60

4. $12\frac{1}{2}$% of $___?___ = $42

5. 8% of $___?___ = $4.96

6. $16\frac{2}{3}$% of $___?___ = $264

7. $37\frac{1}{2}$% of $___?___ = 18

8. $\frac{1}{2}$% of $___?___ = $10

9. $62\frac{1}{2}$% of $___?___ = $55

10. $8\frac{1}{3}$% of $___?___ = $9

1. _____
2. _____
3. _____
4. _____
5. _____
6. _____
7. _____
8. _____
9. _____
10. _____

Problems

Find the rate, to the nearest hundredth, where necessary, in each of the following problems.

Answers

11. $18 is what percent of $72?   **12.** What percent of 128 is 3.84?

11. ——————————

12. ——————————

13. ——————————

14. ——————————

15. ——————————

16. ——————————

13. $78 is what percent of $156?   **14.** $13.50 is what percent of $162?

17. ——————————

18. ——————————

19. ——————————

20. ——————————

15. 64 is what percent of 192?   **16.** ———?——— % of $60 is $6

17. ———?——— % of 150 = 30   **18.** ———?——— % of 63 = $7\frac{7}{8}$

19. ———?——— % of $64 = $56   **20.** ———?——— % of $114 = $38

Increase

An **increase problem** is one in which the new amount (percentage) is larger than the original amount (base). Adding 100% (representing the base) to the rate of increase results in a new rate that represents the new amount (percentage). When the base is multiplied by this new rate, the resulting product will be the new percentage.

When the percentage and the rate of increase are known, the base is found by dividing the percentage by 100% plus the rate of increase.

a. The price of coffee has just increased by 20% from $2.15 a pound. What is the increased price?

b. Coffee is now selling at $2.58 per pound, which is 20% greater than its earlier price. What was the price of coffee before the increase?

a.

B	$\times$	R	$=$	P
Original Amount	$\times$	(100% + Rate of Increase)	$=$	New Larger Amount

$$\$2.15 \times (100\% + 20\%) = ?$$
$$\$2.15 \times 1.20 = \$2.58$$

b. The base is unknown. Divide the percentage ($2.58) by the rate (100% + 20% increase) to find the original price.

P	$\div$	R	$=$	B
New Larger Amount	$\div$	(100% + Rate of Increase)	$=$	Original Amount

$$\$2.58 \div (100\% + 20\%) = ?$$
$$\$2.58 \div 1.20 = \$2.15$$

Check

$$\$2.15 \times 0.20 = \$0.43 \qquad \$2.15 + \$0.43 = \$2.58$$

Self-Check

a. Find the percentage: $28 increased by 15%

b. Find the base: ___?___ increased by 28% = 96.

Self-Check Answers

a. $32.20

b.   75

Problems

Find the percentage in each problem.

21. 120 increased by 10%   **22.** 56 increased by 25%

23. $6.80 increased by 20%   **24.** 50 increased by 100%

Find the base of each problem.

25. ___?___ × 1.25 = 60   **26.** ___?___ × 1.6 = 80

Answers

21. _____

22. _____

23. _____

24. _____

25. _____

26. _____

Problems

27. ___?___ $\times 1.20 = 240$

27. _____

28. ___?___ $\times 2.00 = 356$

28. _____

29. ___?___ increased by 20% = 120

29. _____

30. ___?___ increased by 60% = 32

30. _____

31. ___?___ increased by $12\frac{1}{2}$ = 99

31. _____

32. ___?___ increased by 100% = 78

32. _____

Calculator Tip

When finding rate or base, use the percent key instead of the equals key to multiply or divide by a percent.

Decrease

With **decrease problems,** the percentage—the new amount—is always less than the base. Subtracting the rate of decrease from 100% (representing the base) results in a new rate. When the base is multiplied by this new rate, the resulting product is the new percentage.

When the percentage and the rate of decrease are known, the base is found by dividing the percentage by 100% minus the rate of decrease.

Problems

EXAMPLE

a. A portable CD player was originally priced at $278. The price is to be reduced by 25%. What will be the new (reduced) price?

b. A portable CD player is sale priced at $208.50, which is a 25% reduction from the original price. What was the original price?

SOLUTION

a.

B	$\times$		R		$=$	P
Original Amount	$\times$	$(100\%$	$-$	Rate of Decrease)	$=$	New Lesser Amount
$278	$\times$	$(100\%$	$-$	$25\%)$	$=$	?
$278	$\times$		0.75		$=$	$208.50

b.

P	$\div$		R		$=$	B
New Lesser Amount	$\div$	$(100\%$	$-$	Rate of Decrease)	$=$	Original Amount
$208.50	$\div$	$(100\%$	$-$	$25\%)$	$=$	?
$208.50	$\div$		0.75		$=$	$278

Check

$278 \times 0.25 = \$69.50$ $278 - \$69.50 = \$208.50$

▶ ***Self-Check***

a. Find the percentage: 270 reduced by 40%.
b. Find the base: ___?___ decreased by 24% = 2,280.

Self-Check Answer

a. ___162___

b. ___3,000___

Find the percentage in each of the following decrease problems. Round percent to four decimal places.

Answers

33. 40 decreased by 25%

34. $120 reduced by $33\frac{1}{3}\%$

35. $480 reduced by 16%

36. 978 reduced by 50%

37. 58 reduced by $37\frac{1}{2}\%$

38. $129 reduced by $16\frac{2}{3}\%$

33. _____
34. _____
35. _____
36. _____
37. _____
38. _____

Problems

Find the base in each decrease problem. Round percents to four decimal places.

Answers

39. _____?_____ × 0.78 = 78

40. _____?_____ × 0.625 = 640

41. _____?_____ × 0.45 = 33.75

42. _____?_____ × 0.50 = 1,025

43. _____?_____ decreased by 29% = $177.50

44. _____?_____ decreased by 15% = 85

45. _____?_____ decreased by $8\frac{1}{3}\%$ = $44

46. _____?_____ decreased by 60% = 48

47. _____?_____ decreased by $33\frac{1}{3}\%$ = 950

48. _____?_____ decreased by 62.5% = 642

39. _____

40. _____

41. _____

42. _____

43. _____

44. _____

45. _____

46. _____

47. _____

48. _____

Business Applications

In the following problems, round percents to two decimal places where necessary.

Answers

49. Barthow's Pharmacy sells Plus Screen Lotion for $4 after a 20% discount. What was the original price?

49. _____

50. _____

51. _____

50. Orbit Sales mistakenly included 5% sales tax in a bill to a tax-exempt organization. The invoice showed a total of $247.80, which included the 5% sales tax. What is the amount of tax that must be refunded?

51. Membership in the Four Seasons Raquet Club has dropped by $12\frac{1}{2}\%$, according to Natashia Ling, the manager. This year's membership is 2,198 adults and children. What was last year's membership?

Student Notes

SKILLBUILDER 5.3

Reading Tables

Learning Outcomes

After completing this skillbuilder, you should be able to:

■ Use tables to find information needed to solve problems.

Using Tables

A **table** is a systematic arrangement of information, usually numeric. The data are arranged in labeled rows and columns. Tables are read by selecting the row and column containing the information sought. The data required are found at the intersection of the row and the column. Tables are often used in business to provide a clear and concise presentation of data.

The table on the next page presents population data for 40 of the 150 largest metropolitan statistical areas (MSA). Past data show the percent of population change from 1970 to 1980; actual population in 1980; 1980–1991 estimated percent of change in population; 1991 estimated population; 1991–1996 projected percent of change in population; and 1996 projected population. The MSAs are ranked by population according to their 1980 actual, 1991 estimated, and 1996 projected figures.

	Past			Present			Future		
	1970–1980 %	1980 Census	1980 Rank	1980–1991 %	1991 Estimate	1991 Rank	1991–1996 %	1996 Projection	1996 Rank
Los Angeles–Long Beach	6.3	7,477,503	2	21.5	9,087,000	1	6.3	9,656,500	1
New York	−8.8	8,274,961	1	4.3	8,627,200	2	1.7	8,774,900	2
Chicago	−0.6	6,060,387	3	0.8	6,109,500	3	2.2	6,244,300	3
Philadelphia	−2.1	4,716,818	4	4.4	4,924,100	4	3.5	5,097,100	4
Detroit	−1.4	4,488,072	5	−1.9	4,402,000	5	1.3	4,460,100	5
Washington	7.0	3,250,822	7	24.7	4,055,200	6	8.1	4,383,600	6
Boston–Lawrence– Salem–Lowell– Brockton	−1.2	3,662,832	6	3.4	3,785,800	7	−1.2	3,741,100	7
Houston	44.7	2,735,766	8	23.2	3,370,600	8	5.3	3,549,100	8
Atlanta	26.8	2,138,231	13	40.5	3,003,800	9	12.9	3,391,500	9
Riverside–San Bernardino	36.3	1,558,182	24	81.2	2,823,700	10	16.3	3,284,200	10

MATH TIP

Memorizing commonly used fraction-percent equivalents can be helpful when using tables.

EXAMPLE

Estimate the population of Houston in 1970.

SOLUTION

From 1970 to 1980, the population of Houston grew by 44.7%. Thus, the 1980 population was about 150% of the 1970 population. We can round the 1970 population to 3,000,000.

$$150\% \times 1970 \text{ population} = 3,000,000$$

$$1970 \text{ population} = \frac{3,000,000}{1.5}$$

$$= 2,000,000$$

 **Self-Check**

About how many times as great as the population of Atlanta is the population of Los Angeles?

Self-Check Answer

_____3_____

Problems

1. List three of the four cities whose population ranking remained the same for 1980, 1991, and 1996.

1. _____

_____

_____

_____

2. a. Which two cities dropped in rank by one over the three periods shown?

b. Which two cities rose in rank by one over the three periods shown?

2. **a.** _____

_____

b. _____

_____

3. a. Which city has the greatest change in rank from 1980 to 1996?

b. What is the projected amount of population change for this city from 1980 to 1996?

c. What is the percent of population increase or decrease for this city from 1980 to 1996? (*Note:* The percent of increase or decrease is found by comparing the change in population to the original population.)

3. **a.** _____

_____

b. _____

c. _____

Problems

Food	1985	1986	1987	1988	1989	1990	1991
Chicken, fresh, whole	0.78	0.87	0.74	0.89	0.88	0.86	0.86
Chicken breast	1.74	1.99	1.72	2.06	2.01	2.00	2.02
Turkey, frozen	1.03	1.02	0.89	0.97	0.95	0.96	0.91
Tuna, canned	2.04	1.94	2.11	2.18	2.04	2.11	2.05
Eggs, Grade A, large (doz)	0.91	0.91	0.73	0.83	1.14	1.00	1.01
Milk, fresh, whole ($\frac{1}{2}$ gal)	1.11	1.13	1.16	1.21	1.37	1.39	1.40
Ice cream ($\frac{1}{2}$ gal)	2.29	2.41	2.44	2.54	2.67	2.54	2.63
Apples, Red Delicious	0.68	0.66	0.55	0.71	0.57	0.77	0.86
Bananas	0.32	0.33	0.38	0.41	0.42	0.43	0.42
Pears, Anjou	0.69	0.72	0.57	0.64	0.76	0.79	0.88
Potatoes, white	0.17	0.26	0.25	0.30	0.31	0.32	0.28
Lettuce, iceberg	0.71	0.56	1.25	0.77	0.52	0.58	0.69
Tomatoes, field grown	0.95	0.94	1.00	0.81	0.90	0.86	0.79
Peanut butter	1.57	1.68	1.82	1.82	1.82	2.07	2.04
Sugar, white	0.35	0.35	0.36	0.38	0.41	0.43	0.42

Source: U.S. Bureau of Labor Statistics, *CPI Detailed Report,* January issues.

Refer to the table above showing retail prices.

Answers

4. a. Which three food items had lower prices in 1991 compared to 1985?

b. What was the percent of change in the price of ice cream from 1989 to 1990? (*Note:* Percent of change = $\frac{\text{change in price}}{\text{old price}} \times 100\%$)

c. What was the percent of change in the price of ice cream from 1988 to 1990?

d. What was the trend in the price of whole chickens from 1988 to 1991?

4. a. _____

_____

_____

b. _____

c. _____

d. _____

Refer to the table on the next page showing school enrollments.

5. a. Name two states that did not project an increase in grades 9–12 enrollment from 1990 to 1992.

b. Approximately what percent of all 50 states projected an overall increase in enrollment in grades K–12 from 1990 to 1992?

5. a. _____

_____

b. _____

_____

Problems

State	K Through Grade 8			Grades 9 Through 12			State	K Through Grade 8			Grades 9 Through 12		
	1990	1991	1992	1990	1991	1992		1990	1991	1992	1990	1991	1992
U.S.	29,742	30,186	30,663	11,284	11,389	11,587	MO	588	594	601	222	232	237
AL	529	531	533	199	195	197	MT	109	108	108	43	43	42
AK	82	84	84	30	28	29	NE	197	197	196	77	77	78
AZ	475	495	513	115	165	172	NV	145	151	158	52	54	56
AR	314	315	316	121	123	125	NH	130	136	141	41	46	48
CA	3,611	3,728	3,850	1,352	1,373	1,410	NJ	783	800	824	300	300	302
CO	414	421	428	155	155	159	NM	208	212	216	92	93	95
CT	349	357	367	120	120	120	NY	1,813	1,831	1,854	750	768	779
DE	74	76	76	26	27	29	NC	782	796	815	301	302	306
DC	61	61	62	19	20	19	ND	85	84	83	32	33	34
FL	1,364	1,426	1,488	498	467	479	OH	1,256	1,262	1,268	514	514	519
GA	854	875	898	298	304	310	OK	422	418	415	157	156	158
HI	126	129	133	45	47	49	OR	342	345	347	143	137	141
ID	155	155	152	66	60	62	PA	1,174	1,188	1,201	494	493	500
IL	1,288	1,294	1,305	515	517	521	RI	101	102	105	37	37	37
IN	681	682	682	275	280	285	SC	451	456	462	171	171	174
IA	339	336	330	145	141	145	SD	95	96	96	34	35	36
KS	319	322	323	117	120	125	TN	597	602	608	225	225	227
KY	451	445	440	179	177	181	TX	2,468	2,493	2,516	885	887	892
LA	577	573	570	202	194	197	UT	322	319	316	123	126	133
ME	154	157	160	62	60	60	VT	71	71	72	25	26	26
MD	529	549	571	186	190	196	VA	733	755	782	265	269	274
MA	606	621	637	223	225	226	WA	607	620	632	225	232	240
MI	1,145	1,153	1,161	432	436	440	WV	223	216	210	100	97	95
MN	543	552	558	209	215	222	WI	563	567	569	228	235	241
MS	366	364	362	134	132	135	WY	69	67	66	29	27	27

Source: U.S. National Center for Education Statistics, *Projections of Education Statistics,* annual.

Problems

The following table shows per capita spending on libraries in the United States.

FUNDING OF PUBLIC LIBRARIES, 1991					
United States	$17.83	Kentucky	$ 9.16	North Dakota	$ 8.70
Alabama	$ 9.88	Louisiana	$12.62	Ohio	$28.34
Alaska	$28.64	Maine	$15.40	Oklahoma	$12.37
Arizona	$16.16	Maryland	$25.52	Oregon	$17.88
Arkansas	$ 6.75	Massachusetts	$21.36	Pennsylvania	$12.92
California	$17.54	Michigan	$15.51	Rhode Island	$18.63
Colorado	$20.51	Minnesota	$20.58	South Carolina	$9.67
Connecticut	$27.20	Mississippi	$7.62	South Dakota	$13.59
Delaware	$ 9.48	Missouri	$15.50	Tennessee	$ 7.44
District of Columbia	$35.62	Montana	$ 9.02	Texas	$ 9.53
Florida	$15.09	Nebraska	$15.38	Utah	$15.71
Georgia	$11.62	Nevada	$16.41	Vermont	$13.24
Hawaii	$19.39	New Hampshire	$17.40	Virginia	$18.79
Idaho	$13.62	New Jersey	$27.40	Washington	$23.79
Illinois	$22.72	New Mexico	$10.71	West Virginia	$ 8.38
Indiana	$22.80	New York	$30.36	Wisconsin	$19.10
Iowa	$14.67	North Carolina	$12.01	Wyoming	$22.40
Kansas	$17.41				

Source: National Center for Education Statistics, U.S. Dept. of Education; per capita spending.

Calculator Tip

If your calculator does not have a percent key, change the percent to its decimal equivalent to find a percentage.

6. a. Which state spends the most on libraries, per person?
   b. The 1991 population in Tennessee was about 5 million. About how much did Tennessee spend in all on libraries?

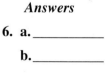

Answers

6. a. _____

b. _____

_____

Business Applications

7. The owner of a chain of fast-food restaurants made the following table of population figures to help decide where in Idaho might be a good place to open her first Idaho store.

City	1980 Population	1990 Population	Percent of Change
Boise	102,249	125,551	_____
Caldwell	17,699	18,400	_____
Cour D'Alene	19,913	24,563	_____
Idaho Falls	39,739	43,929	_____
Meridian	27,986	28,082	_____
Moscow	6,658	9,596	_____
Nampa	25,112	28,365	_____
Pocatello	46,340	46,117	_____
Twin Falls	26,209	27,591	_____

a. Complete the table. *Note:* Percent of change =
$$\frac{1990 \text{ population} - 1980 \text{ population}}{1980 \text{ population}} \times 100\%$$

b. Which city shows the highest percent of population growth?

c. Based on population alone, which city might the owner not consider?

d. What other factors might she consider other than population growth when deciding where to locate?

Answers

7. **b.** _____

c. _____

d. _____

_____

_____

_____

_____

Software buyout

Novell said this week it will buy WordPerfect and Borland's Quattro Pro, making it one of the USA's biggest software marketers. The top-selling software programs:

Software	Type	'93 revenue (millions)
1. Microsoft Word	Word processing	$923
2. Microsoft Excel	Spreadsheet	$779
3. WordPerfect	Word processing	$717
4. Lotus 1-2-3	Spreadsheet	$585
5. Borland Paradox	Database	$207
9. Borland Quattro Pro	Spreadsheet	$108

Source: InfoCorp MarketModel   By Elys A. McLean, USA TODAY

8. What were the total 1993 sales for the two Borland software packages? The Paradox expects a 15% increase and the Quattro Pro expects a 20% increase next year. What will be the expected 1994 sales for each package?

Student Notes

_____
_____
_____
_____
_____
_____
_____
_____
_____
_____
_____
_____
_____
_____

SKILLBUILDER 5.4

Reading Graphs

Learning Outcomes

After completing this skillbuilder, you should be able to:

- Read and interpret line graphs.
- Read and interpret bar graphs.
- Read and interpret circle graphs.

Line Graphs

A **graph** is a visual means of displaying numerical information or data. Graphs are generally less accurate than tables, but they can be used effectively to show trends or comparisons. A **line graph,** or **polygon,** is a form of graph in which information is plotted as points on horizontal and vertical scales. These points are then connected with a broken line. More than one set of data can be shown on a single line graph.

The line graph on the next page shows the sales in one week for three shifts of the clothing department of Dart Round-the-Clock Stores. This graph was constructed by locating, or plotting, the points representing sales for each day (the horizontal scale) on the vertical scale. Each shift was plotted separately, and the points for each shift were then connected. A different type of line was used for each shift.

Problems

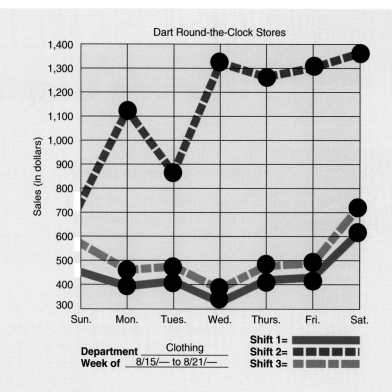

Dart Round-the-Clock Stores

Sales (in dollars)

1,400
1,300
1,200
1,100
1,000
900
800
700
600
500
400
300

Sun. Mon. Tues. Wed. Thurs. Fri. Sat.

Department _Clothing_
Week of _8/15/— to 8/21/—_

Shift 1=
Shift 2=
Shift 3=

EXAMPLE

Determine the approximate sales total on Monday for shift 1.

SOLUTION

$400

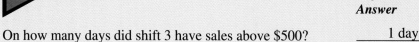

► **Self-Check**

On how many days did shift 3 have sales above $500?

Self-Check Answer

_____1 day_____

MATH TIP

The percentage formula is often used to solve problems involving graphs.

Problems

Answer the following questions using the information shown on the line graph.

1. a. What was the approximate sales total on Wednesday for shift 2?

   b. What was the approximate difference in sales between shift 1 and shift 3 on Saturday?

   c. Which shift had the greatest sales total on Tuesday?

   d. For how many days did shift 2 have sales above $1,000?

   e. Which shift generally has the greatest sales total?

   f. Which shift generally has the least sales total?

Answers

1. **a.** _____

   b. _____

   c. _____

   d. _____

   e. _____

   f. _____

Bar Graphs

A **histogram** is a graph that is useful for showing differences between categories. It consists of perpendicular scales, with bars drawn to represent the amounts. The Sportsway Athletic Supply Store histogram shows the results of a survey of the ages of the first 50 customers to make a purchase at the store on a single day. The horizontal scale gives seven age group ranges. The vertical scale gives the number of customers in each age group. The results of the survey are illustrated by the bars drawn on the graph. The top of each bar indicates the number of persons in each age group.

A **bar graph** is similar to a histogram, except that the bars are separated by space. This form of graph is often used to compare data from one period to another. Bar graphs can be drawn either horizontally or vertically and can present data from more than one period. Bar graphs are read in the same manner as histograms. The Dart Round-the-Clock Stores bar graph compares sales this year with sales for last year in the stores' six departments.

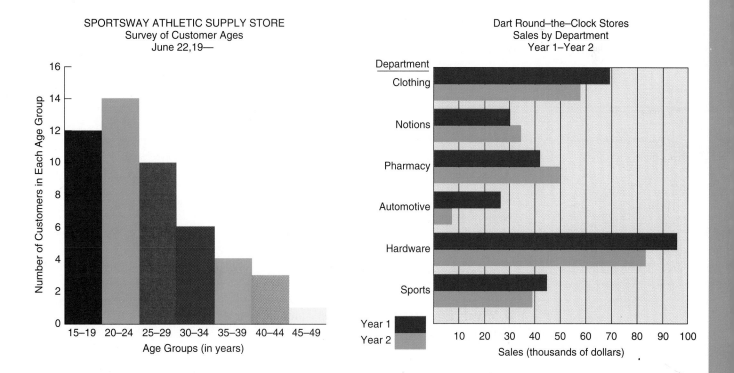

SPORTSWAY ATHLETIC SUPPLY STORE
Survey of Customer Ages
June 22,19—

Dart Round–the–Clock Stores
Sales by Department
Year 1–Year 2

Problems

Use the bar graphs for these problems.

Answers

2. Use the histogram to answer the following questions.

2. a. _____

a. Which age group has 10 customers?

b. _____

b. How many customers are in the 35–39 year age group?

c. _____

c. Which age group had the most customers?

d. _____

d. Which age group had the fewest customers?

e. _____

e. How many customers were between 30 and 49 years of age?

3. Use the bar graph to answer the following questions.

3. a. _____

a. What was the approximate amount of sales in the Notions Department for year 1?

b. _____

b. Which departments had sales over $50,000 in year 1?

c. _____

c. Approximately how much did automotive sales decrease from year 1 to year 2?

d. Which departments had higher sales in year 1 than in year 2?

d. _____

e. _____

e. What is the approximate amount of the total sales in year 1 for the Clothing and Sports Departments combined?

Problems

Circle Graphs

A **circle graph** is an effective way to show how individual parts relate to the whole. For example, in the following circle graph, the entire circle represents $1 of federal government income, or 100%. The total income is divided into different categories to indicate the sources of the income: personal income tax, social security tax and medicare tax (FICA), corporate income tax, and so on. Each of these categories is displayed on a portion of the circle that represents that portion of the income. For example, personal income tax accounts for 35% of every dollar of income.

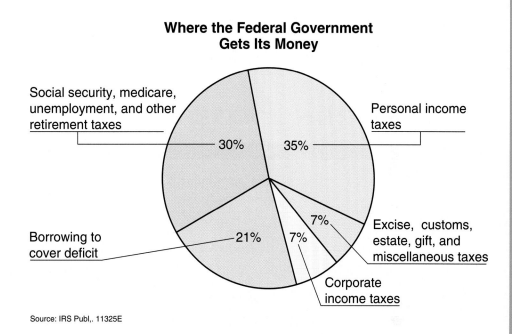

Where the Federal Government Gets Its Money

Social security, medicare, unemployment, and other retirement taxes — 30%

Personal income taxes — 35%

Borrowing to cover deficit — 21%

Corporate income taxes — 7%

Excise, customs, estate, gift, and miscellaneous taxes — 7%

Source: IRS Publ,. 11325E

Calculator Tip

Do not use the percent key to multiply when you use data from a circle graph to find the part of the circle represented by one sector as a decimal.

EXAMPLE

How many cents of every dollar come from personal income tax?

SOLUTION

35% comes from personal income tax
35% of $1 = 0.35 × $1.00 = $0.35

 **Self-Check**

What percent of federal income is obtained from borrowing and corporate income tax together?

Self-Check Answer

_____28%_____

Problems

Answer the following questions using the circle graph on page 211.

4. a. What percent of every dollar of federal income comes from corporate income tax?

b. How many cents of every dollar of federal income come from corporate income tax?

c. What percent of federal income is obtained from social security, medicare, unemployment, and other retirement taxes?

d. How many cents of every dollar of federal income is borrowed to cover the deficit?

e. What percent of federal income is obtained from taxes of all types?

Business Applications

Answer the following questions using the following circle graph.

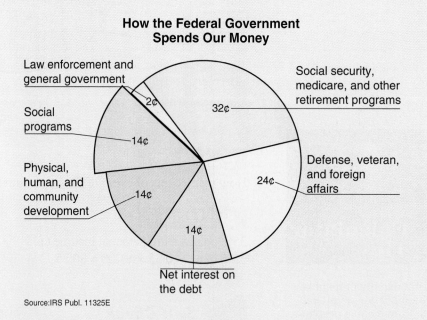

How the Federal Government Spends Our Money

Source: IRS Publ. 11325E

5. a. What percent of the federal budget is spent on interest payments?

b. How many cents of every budget dollar are spent on retirement-related programs?

c. What percent of the federal budget is spent on retirement- and defense-related programs?

d. What percent of the federal budget is spent on the government's general operating costs and law enforcement?

Business Applications

Answer the questions on page 214 using the following circle graphs.

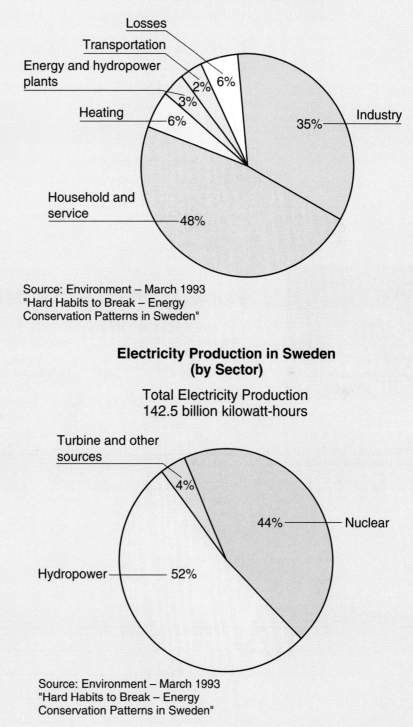

Electricity Consumption in Sweden (by Sector)

Total Electricity Consumption
142.5 billion kilowatt-hours

- Losses — 6%
- Transportation — 2%
- Energy and hydropower plants — 3%
- Heating — 6%
- Industry — 35%
- Household and service — 48%

Source: Environment – March 1993
"Hard Habits to Break – Energy Conservation Patterns in Sweden"

Electricity Production in Sweden (by Sector)

Total Electricity Production
142.5 billion kilowatt-hours

- Turbine and other sources — 4%
- Nuclear — 44%
- Hydropower — 52%

Source: Environment – March 1993
"Hard Habits to Break – Energy Conservation Patterns in Sweden"

6. a. Which sector consumes the greatest amount of electricity?

b. Which sector consumes the least amount of electricity?

c. What percent of the electricity consumed is used in producing electric power?

d. Of the total 142.5 billion kilowatt-hours of electricity consumed, how many billion kilowatt-hours of electricity are lost from the system?

e. Which sector produces the greatest amount of electricity?

f. Which sector produces the least amount of electricity?

g. Of the total quantity of electricity produced, how many billion kilowatt-hours are produced using water-powered sources?

7. Look at the chart below.
a. Which year had the largest deals?
b. Which year had the smallest deals?

Answers

6. a. _____

b. _____

c. _____

d. _____

e. _____

f. _____

g. _____

7. a. _____

b. _____

Calculator Tip

Do not use the percent key to multiply when you use data from a circle graph to find the part of the circle represented by one sector as a decimal.

Hooking up cable TV deals

Takeovers and mergers of cable companies exploded in the late 1980s, until banks clamped down on highly leveraged transactions. In 1993, regional Bell phone companies swept in with new mergers.

Industry deals (in billions)

$16
$12
$8
$4
$0

'84   '87   '90   '93

$10.0

Source: Waller Capital   By Marcy E. Mullins, USA TODAY

Student Notes

SKILLBUILDER 5.5

Computing Sales Commissions

Learning Outcomes

After completing this skillbuilder, you should be able to:

- Compute earnings on a straight commission basis and on a salary plus commission basis.
- Compute earnings on a variable commission basis.

Commissions

Many people who work in sales are paid a percentage of the sales they make. This percentage is called a **commission.** Some people are paid only a commission, and some are paid a commission plus a salary. Others earn a base salary plus a commission only on the amount that exceeds their specified sales quota.

MATH TIP

When figuring sales commission, total sales is the base and the commission rate is the percent.

EXAMPLE

Sheila Collins works for Environmental Products and receives a 15% commission on all the sales she makes. Find her commission during a week in which her total sales were $9,800.

SOLUTION

Commission = sales × commission rate

$9,800   Sales
× 0.15   Commission rate
_____
$1,470   Commission

Sheila's commission is $1,470.

Problems

EXAMPLE

Jamil Toubassi sells furniture for Emmett Designs. He is paid $200 per week plus a 5% commission on all sales. Last week Jamil sold $9,600 worth of furniture. Find his earnings.

SOLUTION

$9,600	Sales		$480	Commission
× 0.05	Commission rate		+ 200	Salary
$480	Commission		$680	Total earnings

Jamil's earnings are $680.

EXAMPLE

Tonya Gibson earns a base salary of $250 per week and an 8% commission on all sales over $2,500. Tonya's sales for the past week were $10,800. Find her earnings for the week.

SOLUTION

$10,800	Sales for week		$8,300	
− 2,500	Minimum quota		× 0.08	Commission rate
$8,300	Sales over $2,500		$664	Commission

$250	Base salary
+ 664	Commission
$914	Total earnings

Tonya's earnings are $914.

 **Self-Check**

Emilio Sanchez works at The Outdoor Store. He receives $250 per week plus a commission of 4% on all sales. Find his earnings in a week when his sales were $6,280.

Self-Check Answer

$501.20

Problems

1. Find the commissions for these salespeople based on their weekly sales.

Salesperson	Rate of Commission	Weekly Sales	Commission
a. Martin Beck	5%	$12,000	————————
b. Jack Longacre	6%	$11,700	————————
c. Jane Hayes	4%	$14,400	————————
d. Alice Bittmore	7%	$10,650	————————
e. Carl Covell	8%	$12,810	————————

2. Find the total of the base salary plus commission for each salesperson based on weekly sales.

Salesperson	Base Salary	Rate of Commission	Weekly Sales	Salary Plus Commission
a. John Bean	$150	5%	$8,250	————————
b. Randy Beasley	$150	6%	6,000	————————
c. Marge Champion	$225	6%	7,600	————————
d. Hilary Dawson	$100	8%	11,100	————————
e. Dawn Gargan	$175	7%	9,750	————————

Problems

3. These salespeople receive a base salary of $175 a week and a commission of 6% on all sales over $1,000. Compute the total earnings of each salesperson.

Salesperson	Weekly Sales	Amount of Commission	Total Earnings
a. Mike Dawes	$ 9,000	_____	_____
b. Pat Hunt	7,580	_____	_____
c. Lois Lipfield	11,700	_____	_____
d. Ruth Moore	21,480	_____	_____
e. Jim Roberts	12,000	_____	_____

Variable Commissions

Some salespeople earn commissions based on a sliding scale. These salespeople may or may not receive a base salary in addition to their commission.

EXAMPLE

Salespeople at Tern Products, Inc., earn a 5% commission on all sales up to and including $2,000: a 7% commission on sales over $2,000 up to and including $4,000; and a 10% commission on all sales over $4,000. Celia Chang had sales of $9,200 one week. What were her total earnings for the week?

Sales	×	Rate	=	Commission
$2,000	×	0.05	=	$100
2,000	×	0.07	=	140
5,200	×	0.10	=	520
		Total	=	$760

Celia's earnings are $760.

 **Self-Check**

Bryan Tarabochia works at Tern Products, Inc. Find his commission on sales of $10,100.

Self-Check Answer

_____$850_____

© by Glencoe/McGraw-Hill.

Problems

4. Compute the total earnings for each of the following sales-people. The commission paid is based on the sliding scale at Tern Products, Inc.

Salesperson	Total Sales	Amount of Commission	Base Salary	Total Earnings
a. Bassett, M.	$ 8,500	———————	$75	———————
b. Copeland, J.	10,000	———————	None	———————
c. DelTorro, V.	11,000	———————	100	———————
d. Englert, T.	9,200	———————	125	———————
e. Franklin, H.	12,200	———————	100	———————
f. Fusco, C.	11,600	———————	None	———————
g. Handley, E.	10,280	———————	125	———————
h. Montez, W.	10,960	———————	100	———————
i. Oroz, J.	12,240	———————	125	———————
j. Pappas, N.	15,200	———————	None	———————
k. Preston, O.	13,600	———————	125	———————
l. Restic, V.	8,800	———————	100	———————
m. Savage, K.	12,400	———————	100	———————
n. Truax, P.	9,880	———————	75	———————
o. Weston, A.	12,800	———————	100	———————
p. Zeramba, L.	6,880	———————	100	———————
	Totals	———————		———————

Business Applications

5. Maria Santangelo is a sales representative for Blitz Lasers. She is compensated for sales (less returns or cancellations) on the following basis:

0–$7,500	→	$3\frac{1}{2}\%$
7,501.01–40,000	→	5%
40,000.01–75,000	→	7%
75,000.01 and over	→	10%

Maria's sales for the period were $105,000. A $17,000 order from the previous sales period was canceled by the customer. How much was her commission for the present period?

6. Pedro Chang is a retail salesperson who is paid an hourly rate of $8.53, with overtime at time and a half for any hours in excess of 40 per week. In addition, he also receives a $7\frac{1}{2}\%$ commission on all sales over $2,500 per week. Pedro worked a total of 53 hours last week and had sales totaling $8,659. What were his gross earnings for the week?

Answers

5. _____

6. _____

Business Applications

7. Christine Yen is paid a commission of $8\frac{3}{8}\%$ on all sales, excluding shipping charges. Last month Christine had sales of $12,973, including shipping charges of $324.33; $4,128, including shipping charges of $103.20; $29,697, including shipping charges of $742.43; and $51,097, including shipping charges of $1,277.43. What was her total commission for the month?

7. _____

8. Amy Snow works at Computers Inc. and earns a 10.5% commission on all software sales. In May, Amy sold $7,352 in software. Compute Amy's commission.

8. _____

Computing personal finances

Personal finance software revenue grew more than twice as fast as total software revenue in 1993 vs. 1992.

U.S. revenue

19%
All software

51%
Personal finance software

Source: Software Publishers Association

By Cindy Hall and Marcy E. Mullins, USA TODAY

Student Notes

SKILLBUILDER 5.6

Computing Agents' and Brokers' Commissions

Learning Outcomes

After completing this skillbuilder, you should be able to:

■ Compute earnings for agents and brokers.

Agent and Broker Commissions

Agents and brokers are individuals who work for themselves but who represent the business interests of others. They are usually compensated on a commission basis. A sales representative who is employed by a company may be paid a salary plus a commission and is usually reimbursed for expenses incurred in selling. Agents or brokers may represent manufacturers, importers, exporters, athletes, artists, or property owners. The contract between the representative and the principal (client) determines the basis for compensation.

MATH TIP

Remember, a misplaced decimal point will cause an error in the amount of commission.

EXAMPLE

Peter Pfaffenroth is an automobile broker who receives $7\frac{1}{2}\%$ commission on all sales he concludes. How much commission will he receive on the sale of a 1927 Ford for $13,275?

SOLUTION

$Commission = Sales \times Rate\ of\ Commission$

$\$995.63 = \$13,275 \times 0.075$

Peter's commission is $995.63.

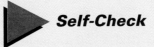 **Self-Check**

Pavi Terhar is a real estate broker. She receives a 7% commission on the sale of a house. How much commission does she receive on a house that sells for $98,000?

Self-Check Answer

$6,860

Problems

Langford Boutique pays its employees $5 per hour plus a commission of 2.5% of all sales. Compute the wages for each employee.

Name	Hours Worked	Sales	Wages
1. Goldsmith, A.	23	$1,320	_____
2. Palusinska, K.	40	$2,544	_____
3. Trinh, P.	32	$2,065	_____
4. Varga, J.	33	$1,955	_____
5. Zurlinden, M.	40	$2,200	_____

Solve each problem.

6. Arlene Jablonski had sales of $250,000 for the months of January through March. She is paid a commission of 8% on all sales. What was her commission for the quarter?

7. Pedro Chang is paid 10% commission on sales to retailers and 6% commission on sales to wholesalers. If sales to retailers were $90,000 and sales to wholesalers were $175,000, what was the total commission he earned?

8. Nicole Koshak is paid 12% commission on net sales (sales minus returns). Sales for the period were $50,000, with returns of $1,800. What was her commission?

9. Harold Huizenga is paid $4.50 an hour plus 3% of all sales. How much did he earn in a week in which he worked 32 hours and sold $6,300?

Answers

6. _____

7. _____

8. _____

9. _____

Calculator Tip

Enter the rate of commission as a decimal. Press the ⊗ key, the sales amount, and the ⊜ key to calculate the amount of commission.

Business Applications

10. CheJan Park, an importer's agent, sold a total of $185,000 of imported manufactured products to various importers. Che-Jan receives a commission a $4\frac{3}{4}\%$ on all orders sold. How much did she earn on this group of sales?

10. _____

11. Bernadette Pavley, an artist's agent, receives 7% commission on all concert fees paid her artists, 3% commission on all records sold at public appearances, and $1\frac{1}{2}\%$ commission on all product endorsements. Total concert fees for the year were $3,679,000; record sales totaled $273,000; product endorsement contracts totaled $1,936,852. What was Bernadette's gross income for the year?

11. _____

12. Marion Castello, a real estate sales agent, sold a commercial property for $875,000. Her broker receives an 8% commission on the sale, and Marion receives 30% of the total commission. How much is Marion's commission on this sale?

12. _____

13. A leasing agent receives a 12% commission on each contract she completes. If the agent handled 10% of the total customers in 1992, compute the commissions paid to her that year.

Answers

13. _____

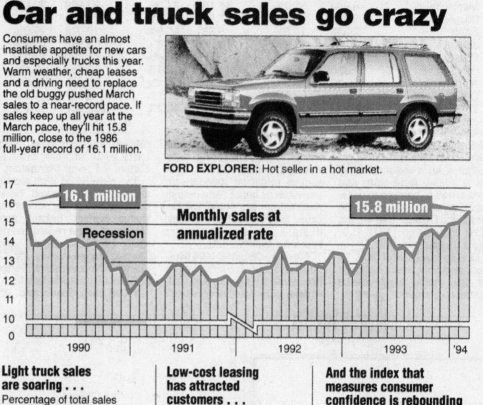

Car and truck sales go crazy

Consumers have an almost insatiable appetite for new cars and especially trucks this year. Warm weather, cheap leases and a driving need to replace the old buggy pushed March sales to a near-record pace. If sales keep up all year at the March pace, they'll hit 15.8 million, close to the 1986 full-year record of 16.1 million.

FORD EXPLORER: Hot seller in a hot market.

16.1 million Recession

Monthly sales at annualized rate

15.8 million

1990   1991   1992   1993   '94

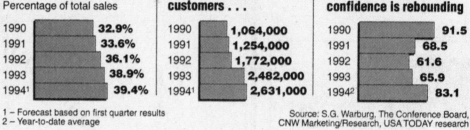

Light truck sales are soaring . . . Percentage of total sales		Low-cost leasing has attracted customers . . .		And the index that measures consumer confidence is rebounding	
1990	32.9%	1990	1,064,000	1990	91.5
1991	33.6%	1991	1,254,000	1991	68.5
1992	36.1%	1992	1,772,000	1992	61.6
1993	38.9%	1993	2,482,000	1993	65.9
1994[1]	39.4%	1994[1]	2,631,000	1994[2]	83.1

1 – Forecast based on first quarter results
2 – Year-to-date average

Source: S.G. Warburg, The Conference Board, CNW Marketing/Research, USA TODAY research

By Marty Baumann, USA TODAY

Student Notes

_____

_____

_____

_____

SKILLBUILDER 5.7

Computing Markup

Learning Outcomes

After completing this skillbuilder, you should be able to:

- Compute markup and percent of markup based on cost price.
- Compute markup and percent of markup based on selling price.
- Find the markup and selling price.

Markup Based on Cost Price

Retailers make a profit and cover their expenses by charging more for a product than they pay for it. The amount added to the cost price is the **markup.** The markup rate is called the **percent of markup,** which can be based on the cost price or on the selling price.

Most retail merchants use the rate of markup based on cost to determine their selling price. For example, if a retail merchant decides that all items must have a 25% markup based on cost, an item costing $24 would be sold for $30.

$$\$24 \times 0.25 = \$6 \qquad \$24 + \$6 = \$30$$

If the same merchant decides to use a 50% markup based on cost, the item costing $24 would sell for $36.

$$\$24 \times 0.50 = \$12 \qquad \$24 + \$12 = \$36$$

MATH TIP

Remember to enter the decimal in the selling price and to change the markup percent to a decimal.

EXAMPLE

An item costs $37.50; its selling price is $43.75. Find (a) the markup and (b) the rate of markup based on the cost price.

SOLUTION

a.
$$\begin{array}{ll} \$43.75 & \text{Selling price} \\ \underline{-37.50} & \text{Cost price} \\ \$6.25 & \text{Markup} \end{array}$$

b. *Rate of Markup Based on Cost Price = Markup ÷ Cost Price*

$$16\tfrac{2}{3}\% \quad \text{or} \quad 16.67\% = \quad \$6.25 \quad \div \quad \$37.50$$

▶ **Self-Check**

A chair costs $220; its selling price is $330. Find (a) the markup and (b) the rate of markup based on the cost price.

Self-Check Answers

a. $110

b. 50%

Problems

Solve each problem.

Answers

1. Julie Carter purchased radios for $18 each and sold them for $24.
a. Find the markup.
b. Find the rate of markup, to the nearest hundredth of a percent, based on cost price.

1. a._____

   b._____

2. Pierre Products purchased electronic typewriters for $150 each and sold them for $191.60.
a. Find the markup.
b. Find the rate of markup, to the nearest hundredth of a percent, based on cost price.

2. a._____

   b._____

3. Caryl Clothes purchased a dozen blouses at $216 a dozen. Each blouse retailed for $23.31. What was the rate of markup, to the nearest tenth of a percent, based on cost price?

3. _____

4. Goliath Corp. purchased a gross of calculators for $3,730. (There are 144 items in a gross.) The calculators will be priced at a markup of $37\frac{1}{2}$ percent of the cost price. What will be the selling price of each calculator?

4. _____

5. The Alberta Shoe Emporium realizes a profit of $13 on each pair of a certain style of shoe that it sells. This represents a 50 percent markup based on the cost price.
a. What is the cost price of each pair of shoes?
b. What is the selling price of each pair?

5. a._____

   b._____

Problems

Markup Based on Selling Price

Markup can also be based on the selling price, or retail price, of the goods.

Calculator Tip

To figure the markup rate on a calculator, key in the markup, press the ÷ key, key in the cost (or selling price), and press the % key.

EXAMPLE

Find the rate of markup based on the selling price for a bicycle that cost $75 and that sells for $87.50.

SOLUTION

$87.50   Selling price
−75.00   Cost price
—————
$12.50   Markup

Rate of Markup
Based on Selling Price = Markup ÷ Selling Price
14.29% = $12.50 ÷ $87.50

 **Self-Check**

A TV costs $201.60. The markup is $86.40. Find the percent of markup based on the selling price.

Self-Check Answer

_____30%_____

Solve each problem.

6. Portable typewriters cost a dealer $144 each. The markup is $66.
a. What is the selling price of each typewriter?
b. What is the rate of markup to the nearest hundredth of a percent based on the selling price?

7. A living room chair cost $270 and sells for $324.
a. What is the amount of the markup?
b. What is the rate of markup to the nearest hundredth of a percent based on the selling price?

Answers

6. a._____

b._____

7. a._____

b._____

Problems

Finding Markup and Selling Price

Cost price divided by cost price expressed as a percent of selling price will give the selling price.

EXAMPLE

A radio cost $37.50 and will be sold at a markup of 25% based on the selling price. Find the markup and the selling price.

SOLUTION

Selling price   100%
Cost price   −25%   of selling price
Markup   75%   of selling price

Selling Price = *Cost Price* ÷ *Percent Cost Price Is of Selling Price*
    $50.00 = $37.50   ÷   75%
    $50.00   Selling price
  −37.50   Cost price
  $12.50   Markup

 **Self-Check**

A calculator cost $48 and will be sold at a markup of $37\frac{1}{2}$% based on the selling price. Find (a) the selling price and (b) the markup.

Self-Check Answer

a. $76.80
b. $28.80

Solve each problem.

8. A suitcase cost $43.56 and was sold at a markup of 50% of the selling price.
a. What was the selling price?
b. What was the amount of markup?

Answers

8. a. _____
   b. _____
9. a. _____
   b. _____

9. A digital alarm clock cost $18 and was sold at a markup of 40% of the selling price.
a. What was the selling price?
b. What was the amount of markup?

Business Applications

10. Complete this table. Round off the rate to the nearest hundredth of a percent, where necessary.

Answers

	Cost Price	Markup	Selling Price	Rate of Markup Based On	
				Cost Price	Selling Price
a.	$86.40	_____	$103.68	_____	_____
b.	46.20	6.60	_____	_____	_____
c.	_____	2.48	_____	16%	_____
d.	_____	4.80	_____	_____	8.75%

11. Fancy Frocks spring line includes designer shorts, which cost them $9.87. Their usual markup is 70% of selling price. What is the selling price?

11. _____

12. Home-Pro Tools sell overstock tools from their outlet stores at an average markup of $37\frac{3}{8}\%$ of cost. What should the selling price be for each of the following?
a. A cordless screwdriver costing $11.85.
b. A $\frac{3}{8}$-inch drill costing $21.50.
c. $7\frac{1}{4}$-inch circular saw costing $31.38.

12. **a.** _____

b. _____

c. _____

13. The Jewelry Chest manufactures custom jewelry to order, which they usually mark up at 250% of production cost. The cost for design, labor, and materials for a gold pendant was $87.50. What was their selling price for this item?

13. _____

14. If a pair of hiking boots which cost Nike $65 are marked up 130%, what will be the selling price?

Answer

14. _____

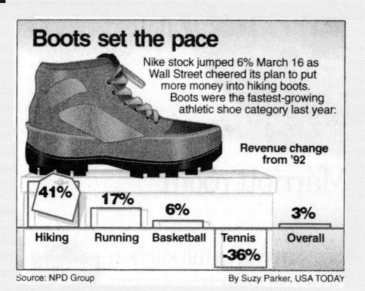

Boots set the pace

Nike stock jumped 6% March 16 as Wall Street cheered its plan to put more money into hiking boots. Boots were the fastest-growing athletic shoe category last year:

Revenue change from '92

Hiking	Running	Basketball	Tennis	Overall
41%	17%	6%	-36%	3%

Source: NPD Group

By Suzy Parker, USA TODAY

Student Notes

SKILLBUILDER 5.8
Computing Markdown

Learning Outcomes

After completing this skillbuilder, you should be able to:

■ Compute the amount and rate of markdown; the reduced, or new, selling price; and the original selling price.
■ Compute the cost price and selling price when goods are sold at a loss.

Markdown

A reduction in selling price, called **markdown,** is used to move slow-selling merchandise. Markdown is expressed either as an amount or as a rate (percent). The rate of markdown may be based on either the original selling price or the new, or reduced, selling price. The dollar differences are the same, but using the reduced selling price gives the impression of a larger markdown.

MATH TIP

To find the new selling price, find the complement of the markdown rate. The complement is found by subtracting the discount from 100%

EXAMPLE

An item that sells for $10 is marked down to $8. What are the amount of markdown and the rates of markdown based on the original selling price and on the new selling price?

SOLUTION

$10   Original selling price
− 8   New selling price
$ 2   Markdown

Rate of Markdown = Markdown ÷ New Selling Price
   0.25 = 25% = $2   ÷ $8

Rate of Markdown = Markdown ÷ Original Selling Price
   0.20 = 20% = $2   ÷ $10

The markdown is $2. The rate of markdown based on the new selling price is 25%. The rate based on the original selling price is 20%.

 **Self-Check**

An item that sells for $25 is marked down to $18. Find the rate of markdown based on the original selling price.

Self-Check Answer

28%

Problems

Solve the following problems.

1. Find the markdown for the merchandise below. Find the rate of markdown to the nearest tenth of a percent, on both original selling price and new selling price.

	Original Selling Price	New Selling Price	Markdown	Rate of Markdown	
				Original Price	New Price
a.	$25.00	$20.00	_____	_____	_____
b.	84.00	70.00	_____	_____	_____
c.	99.00	88.00	_____	_____	_____
d.	527.00	395.25	_____	_____	_____

Answers

2. The Varsity Shop marked down golf clubs from $94.80 to $79. What was the rate of markdown based on the new selling price?

2. _____

3. Billy Bootery marked down shoes from $50 to $40.
a. What was the rate of markdown based on the new selling price?
b. If the shoes were marked down an additional $15, what was the total rate of markdown based on the new selling price?
c. What was the total rate of markdown based on the original selling price?

3. a._____
   b._____
   c._____

4. A bicycle is marked down $30, which is 20% of the original selling price.
a. What was the original selling price?
b. What is the new selling price?

4. a._____
   b._____

Selling at a Loss

Goods are sometimes sold at a loss. When this is done, the new selling price is less than the cost of the goods. When the percent of loss is based on the new selling price, the cost price of the item is always equal to 100 percent plus the percent of loss based on the new selling price.

Problems

EXAMPLE

a. Find the cost price of an item which was sold for $3.60 at a loss of $16\frac{2}{3}$ percent based on the selling price.

b. Find the new selling price of an item on which the loss is $12.32, which is $33\frac{1}{3}\%$ of the new selling price.

SOLUTION

a.
100%	Selling price
$+16\frac{2}{3}\%$	Loss
$116\frac{2}{3}\%$ or 116.67%	Cost price

$$\text{Cost Price} = \frac{\text{New}}{\text{Selling Price}} \times \frac{\text{Percent Cost Price}}{\text{Is of Selling Price}}$$

$$= \$3.60 \times 116\frac{2}{3}\%$$

$$\$4.200 \text{ or } \$4.20 = \$3.60 \times 1.1667$$

b.

$$\frac{\text{New}}{\text{Selling Price}} = \text{Amount of Loss} \div \frac{\text{Loss as a Percent of}}{\text{New Selling Price}}$$

$$\$36.96 = \$12.32 \div 33\frac{1}{3}\%$$

$$\text{Cost} = \text{New Selling Price} + \text{Amount of Loss}$$

$$\$49.28 = \$36.96 + \$12.32$$

▶ **Self-Check**

Find the new selling price of an item that cost $24.64 if it is sold at a loss of $33\frac{1}{3}$ percent based on the new selling price.

Self-Check Answer

$18.48

Problems

Solve the following problems.

5. Complete the following table. Round off the rate of loss to the nearest tenth of a percent, where necessary.

	Cost Price	New Selling Price	Loss	Rate of Loss Based On	
				Cost Price	New Selling Price
a.	$106.80	$83.20			
b.	2.10		0.42		
c.	6.50				20%
d.		13.65			12.5%
e.	11.20			20%	

Calculator Tip

If you are given the rate of markdown, you can use the percent discount key to find the net price.

6. A sewing machine was sold for $72 less than its cost price, which was $37\frac{1}{2}$% of the new selling price.

a. What is the new selling price?
b. What did the sewing machine cost?

Answers

6. **a.** _____
   b. _____

7. A discontinued model typewriter was sold for $45 less than its cost price. The loss on the sale was 25% of the cost price.
a. Find the cost.
b. Find the new selling cost.

7. **a.** _____
   b. _____

Business Applications

8. Green Village Gardens is selling a group of decorative pots at $3.20, which is a 25% loss. Their usual markup is 40% of selling price. What was the original selling price of the pots?

8. _____

9. McCarter Wholesalers is overstocked on kitchen gadget sets normally selling for $25. On this class of merchandise, McCarter has a markup of 15% on cost. To clear the inventory, it is offering the sets at a loss of 15%. What should the new selling price be?

9. _____

Business Applications

10. The Elegant Boutique has marked down designer jeans three separate times, to a final price of $18. The markdowns were $33\frac{1}{3}\%$, 15%, and 25%. What was the original selling price?

Answers

10. _____

11. Ferguson's Books overstocked *In The Kitchen With Rosie,* and needs to lower the price. The book normally sells for $14.95, but Ferguson's marks it down 30%. What is the new selling price?

11. _____

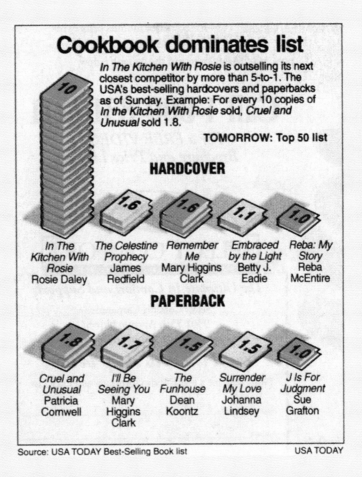

Cookbook dominates list

In The Kitchen With Rosie is outselling its next closest competitor by more than 5-to-1. The USA's best-selling hardcovers and paperbacks as of Sunday. Example: For every 10 copies of *In the Kitchen With Rosie* sold, *Cruel and Unusual* sold 1.8.

TOMORROW: Top 50 list

HARDCOVER

10 — In The Kitchen With Rosie — Rosie Daley
1.6 — The Celestine Prophecy — James Redfield
1.6 — Remember Me — Mary Higgins Clark
1.1 — Embraced by the Light — Betty J. Eadie
1.0 — Reba: My Story — Reba McEntire

PAPERBACK

1.8 — Cruel and Unusual — Patricia Cornwell
1.7 — I'll Be Seeing You — Mary Higgins Clark
1.5 — The Funhouse — Dean Koontz
1.5 — Surrender My Love — Johanna Lindsey
1.0 — J Is For Judgment — Sue Grafton

Source: USA TODAY Best-Selling Book list   USA TODAY

Student Notes

_____

_____

_____

_____

_____

_____

_____

_____

_____

_____

_____

_____

_____

_____

_____

_____

_____

_____

_____

_____

_____

_____

_____

SKILLBUILDER 5.9

Ratios and Proportions

Learning Outcomes

After completing this skillbuilder, you should be able to:

- Determine if two ratios are proportional.
- Solve proportions for the missing value.

Ratios and Proportions

A **ratio** is the relationship of one number to another. A ratio may be written either by separating the numbers by the word *to,* by separating the numbers by a colon (:), or by showing the relationship as a fraction.

For example, if Gardner Electronics employs 30 full-time production workers and 10 part-time production workers, the ratio of full-time to part-time workers is

$$30 \text{ to } 10 \quad or \quad 30{:}10 \quad or \quad \frac{30}{10}$$

This ratio is generally written in simplest form as 3:1. When the fraction form is used, only two numbers can be compared, but when a colon or the word *to* is used, more than two numbers can be compared. For example, if Gardner Electronics also employs five office and clerical workers, the ratio of full-time production workers to part-time production workers to office employees is

$$30 \text{ to } 10 \text{ to } 5 \quad or \quad 30{:}10{:}5 \quad or \quad 6{:}2{:}1$$

A **proportion** is a comparison of two sets of numbers having the same ratio or relationship. In the previous example, the ratio of office workers to total employees is 5 to (30 + 10 + 5), or 5:45. In fraction form, this ratio can be reduced to $\frac{1}{9}$. These two ratios, when compared, form a proportion, as follows:

$$\frac{5}{45} = \frac{1}{9}$$

This expression is read "5 is to 45 as 1 is to 9." Since the ratio, or relationship, is the same between the two sets of numbers, their cross products will be equal. In cross multiplication, the numerator of each ratio is multiplied by the denominator of the other ratio. The resulting products must be equal for the ratios to be a proportion.

$$\frac{5}{45} \bowtie \frac{1}{9} \qquad \begin{aligned} 5 \times 9 &= 45 \\ 1 \times 45 &= 45 \end{aligned}$$

The two ratios are in proportion.

Problems

MATH TIP

Expressions such as *7 out of 8* can be written as 7/8.

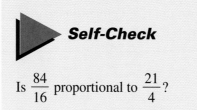

EXAMPLE

Is $\frac{2}{3}$ proportional to $\frac{4}{9}$?

SOLUTION

$$\frac{2}{3} \bowtie \frac{4}{9} \qquad \begin{array}{l} 2 \times 9 = 18 \\ 3 \times 4 = 12 \end{array}$$

Since the cross products are not equal, the ratios are not proportional.

▶ **Self-Check**

Is $\frac{84}{16}$ proportional to $\frac{21}{4}$?

Self-Check Answer

_____ Yes _____

Determine whether the given ratios are proportional.

1. $\frac{5}{9}$, $\frac{20}{36}$

2. $1:1\frac{1}{2}$, $3:4\frac{1}{2}$

3. $\frac{5}{12}$, $\frac{7}{15}$

4. $\frac{3}{8}$, $\frac{9}{24}$

5. $1:2$, $60:180$

Answers

1. _____

2. _____

3. _____

4. _____

5. _____

Solving Proportions

Where two sets of numbers are proportional, but one set is incomplete, the unknown value can be found by cross multiplying.

Problems

Calculator Tip

A ratio suggests division. When solving proportions, use the ⊗ and ⊘ keys.

EXAMPLE

Gardner Electronics is planning an increase in its production line. Three wrapping machines require an area of 28 m². How much space must Gardner provide for a total of 12 wrapping machines?

SOLUTION

Known

Number of machines presently in use—3
Number of machines to be used—12
Area required for 3 machines—28 m²

Unknown

Area required for 12 machines

Ratios

The ratios involved are 12 machines to 3 machines and N m² to 28 m². The proportion would be 12 is to 3 as N is to 28.

1. $\dfrac{12}{3} = \dfrac{N}{28}$

2. Cross multiply 12 by 28: $12 \times 28 = 336$.

3. Place this product over the denominator 3, and carry out the division.

$$N = \frac{336}{3} = 112 \text{ m}^2$$

4. *Check*

$\dfrac{12}{3} = \dfrac{112}{28}$ $12 \times 28 = 336$
                          $3 \times 112 = 336$

Twelve machines would require 112 m² of floor space.

 **Self-Check**

Solve for N: $\dfrac{3}{75} = \dfrac{16}{N}$.

Self-Check Answer

400

Problems

Answers

6. $\dfrac{1}{N} = \dfrac{5}{15}$

6. _____

7. $\dfrac{10.5 \text{ gal}}{250 \text{ mi}} = \dfrac{G \text{ gal}}{300 \text{ mi}}$

7. _____

8. $\dfrac{18}{23} = \dfrac{36}{N}$

8. _____

9. $\dfrac{N}{17} = \dfrac{96}{102}$

9. _____

10. $\dfrac{49}{7} = \dfrac{X}{1}$

10. _____

11. $\dfrac{2}{Y} = \dfrac{10\frac{1}{2}}{420}$

11. _____

12. $\dfrac{H \text{ h}}{25 \text{ mi}} = \dfrac{3 \text{ h}}{5 \text{ mi}}$

12. _____

13. $18:3 = 126:X$

13. _____

14. $\dfrac{4}{56} = \dfrac{C}{784}$

14. _____

15. $8:0.77 = 32:N$

15. _____

Business Applications

16. At Gilray Electronics, it takes 16 production workers to produce 48 parts in 1 h. How many workers would be needed to produce 144 parts in 1 h?

16. _____

17. Fred's Flowers has a new delivery van, which used 25 L of gasoline to travel 290 km. How far can the van travel on 60 L of gasoline?

17. _____

18. Hamburger Heaven can prepare 136 portions from 34 lb of ground beef. How many portions can be prepared from 110 lb of ground beef?

18. _____

19. At Tycee Manufacturing, a quality check found 21 transformers defective out of a production run of 840. Inspection of a second lot found 67 defective units. Assuming a uniform defect rate, how many transformers were in the second lot?

19. _____

20. If 520 people surveyed were comfortable with recording on VCRs and were 50 years old or older, how many people who were comfortable with VCR's were 18–29 years old?

20. _____

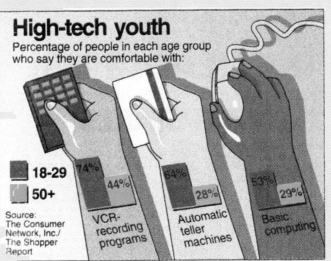

High-tech youth
Percentage of people in each age group who say they are comfortable with:

18-29
50+

74% 44% VCR-recording programs
64% 28% Automatic teller machines
53% 29% Basic computing

Source: The Consumer Network, Inc./ The Shopper Report

By Patti Stang and Elys A. McLean, USA TODAY

U N I T

6

BUSINESS AND CONSUMER MATH

This graph shows factors that people consider when choosing a commercial or savings bank. Why do you think interest rates are more important in choosing a savings bank as opposed to a commercial bank? Why is location more important in choosing a commercial bank?

In this unit we discuss checking accounts and statements, payroll plans, inventory, and depreciation.

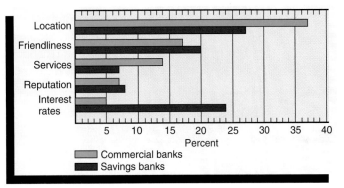

Source: Thompson & Howe, Business Math, Glencoe

Accountant

Judy Jones, a tax accountant, uses her training to gather pertinent information, analyze data, and make recommendations to her clients, based on her knowledge and her understanding of the clients' desires. Judy will then present her findings and recommendations in a readable, understandable format so that her clients can make the best choice. Often tax accountants like Judy recommend that clients open individual retirement accounts (IRAs) as a reasonable way to prepare for retirement and defer federal income tax.

Math Application

Judy's clients, Bob and Jean Holmes, are concerned about retirement but have little extra money to invest. After some discussion, Judy recommends that the Holmes deposit $150 per month in a tax-deferred IRA through Jean's employer. Bob wants to know if this means that Jean will be lowering her take-home pay by $150 each month. Assume the Holmes would normally pay a tax rate of 20% on the $150.

Compute the saved taxes.

$$(20\%)\,(150) = \underline{\hphantom{xxxxx}}$$

Subtract the saved taxes from the $150 to compute the net pay difference.

$$\$150 - \$30 = \underline{\hphantom{xxxxx}}$$

Critical Thinking Problem

The Holmes insist that even $50 a month lower income will cause hardship What might Judy suggest, to help lower the burden of the IRA?

NAME: _____

DATE: _____

SKILLBUILDER 6.1

Checking Accounts and Bank Records

Learning Outcomes

After completing this skillbuilder, you should be able to:

■ Complete a check register or a check stub.

Using Check Registers

A checking account is essential for a business. It allows the business to pay its employees and other bills, as well as to maintain an accurate account of its income and spending.

A **deposit slip** is used to deposit money to an account. The deposit slip shown here indicates that Brenda's Towing Service deposited two checks in the amounts of $85.88 and $195.

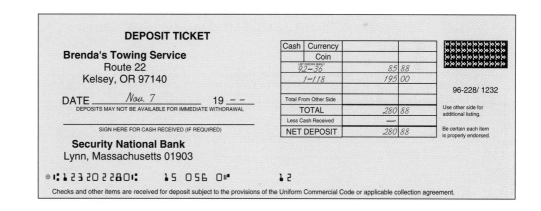

A check is written on the account to pay a bill, for example. When a check is presented to a bank, the bank is authorized to take the amount specified from the account on which the check is drawn. In the check shown here, Brenda's Towing Service has paid $99.58 to Pacific Power for its October electric bill.

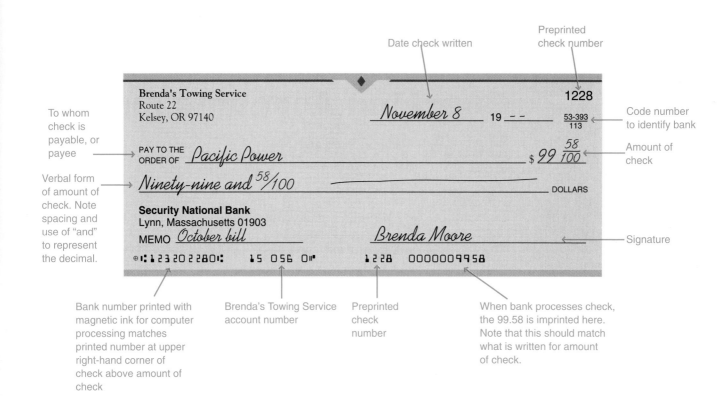

To keep track of deposits made, checks written, and the balance on hand, the depositor uses a **checkbook stub** or **check register.** The stub or register entries should always be filled out before the check is written.

CHECK NO.	DATE	CHECKS ISSUED TO OR DESCRIPTION OF DEPOSIT	AMOUNT OF PAYMENT	✓	AMOUNT OF DEPOSIT	BALANCE FORWARD	
						2,348	56
—	11/7	To Deposit			280 88	Check or Dep. 280	88
		For				Bal. 2,629	44
1228	11/8	To Pacific Power	99 58			Check or Dep. 99	58
		For October electric service				Bal. 2,529	86
—	11/9	To Bank service charge	10 00			Check or Dep. 10	00
		For				Bal. 2,519	86
		To				Check or Dep.	
		For				Bal.	
		To				Check or Dep.	
		For				Bal.	
		To				Check or Dep.	
		For				Bal.	
		To				Check or Dep.	
		For				Bal.	

Whenever a deposit is made, the amount is added to the balance shown in the register.

The amount of any check is deducted from the balance.

Any charges the bank makes for its services should be deducted, and any money collected for the depositor should be added.

It is important to properly identify each transaction and to add or subtract as the nature of the transaction requires.

MATH TIP

Avoid writing checks for less than $1 if possible.

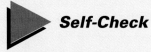

Find the balance in the checking account on the previous page if a check in the amount of $215.90 is written to Cost Club.

SOLUTION

The amount of a check is subtracted:

$$
\begin{array}{r}
\$2{,}519.86 \\
-\ \ \ 215.90 \\
\hline
\$2{,}303.96 \quad \text{Balance}
\end{array}
$$

▶ **Self-Check**

Find the balance in this checking account after checks totaling $213.25 are deposited and a check in the amount of $199.52 is written.

Self-Check Answer

$2,317.69

Problems

Solve these problems.

Answers

1. This check register shows various checks written and deposits made. The balance before check number 576 is $485.14. Assuming all entries are correctly written, what is the balance remaining after check number 581 is written?

1. _____

Calculator Tip

Round all answers to two decimal places unless otherwise instructed.

RECORD ALL CHARGES OR CREDITS THAT AFFECT YOUR ACCOUNT

NO.	DATE	CHECKS ISSUED TO OR DESCRIPTION OF DEPOSIT	AMOUNT OF PAYMENT		✓	OTHER DEDUCT.	AMOUNT OF DEPOSIT	BALANCE FORWARD 485	14
576	6/2	To Connally-Wright / For Insurance	112	75					
577	6/3	To MasterCard / For May statement	116	86					
	6/5	To Deposit / For —					175	00	
578	6/6	To United Campaign / For Contribution	10	00					
579	6/8	To Public Gas & Electric Co. / For May Gas & Electric	75	00					
	6/12	To Deposit / For —					300	00	
580	6/15	To State Motor Vehicles / For License renewal	19	50					
	6/15	To Printing charge / For New check supply	14	31					
581	6/18	To West-Penn Oil Co. / For Fill tank	343	20					
		To / For							

Business Applications

2. The nine check stubs on page 251 are from Portland Pottery. The amount of each check and deposit has been properly recorded. The balance before check number 535 is $1,240.78. What is the balance after check number 543 is written?

2. _____

Business Applications

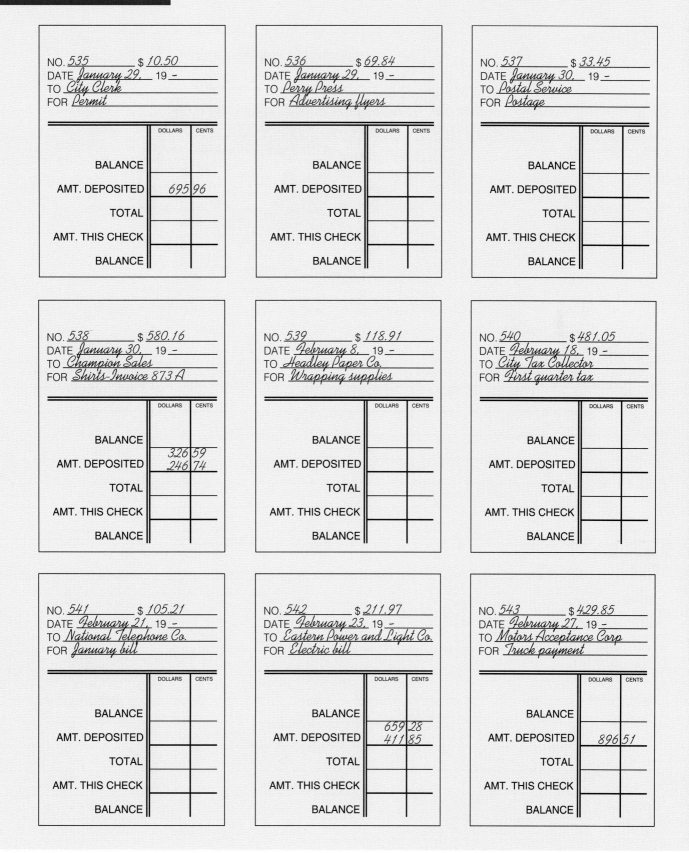

NO. 535 $ 10.50
DATE *January 29,* 19 –
TO *City Clerk*
FOR *Permit*

	DOLLARS	CENTS
BALANCE		
AMT. DEPOSITED	695	96
TOTAL		
AMT. THIS CHECK		
BALANCE		

NO. 536 $ 69.84
DATE *January 29,* 19 –
TO *Perry Press*
FOR *Advertising flyers*

	DOLLARS	CENTS
BALANCE		
AMT. DEPOSITED		
TOTAL		
AMT. THIS CHECK		
BALANCE		

NO. 537 $ 33.45
DATE *January 30,* 19 –
TO *Postal Service*
FOR *Postage*

	DOLLARS	CENTS
BALANCE		
AMT. DEPOSITED		
TOTAL		
AMT. THIS CHECK		
BALANCE		

NO. 538 $ 580.16
DATE *January 30,* 19 –
TO *Champion Sales*
FOR *Shirts-Invoice 873 A*

	DOLLARS	CENTS
BALANCE		
AMT. DEPOSITED	326	59
	246	74
TOTAL		
AMT. THIS CHECK		
BALANCE		

NO. 539 $ 118.91
DATE *February 8,* 19 –
TO *Headley Paper Co.*
FOR *Wrapping supplies*

	DOLLARS	CENTS
BALANCE		
AMT. DEPOSITED		
TOTAL		
AMT. THIS CHECK		
BALANCE		

NO. 540 $ 481.05
DATE *February 18,* 19 –
TO *City Tax Collector*
FOR *First quarter tax*

	DOLLARS	CENTS
BALANCE		
AMT. DEPOSITED		
TOTAL		
AMT. THIS CHECK		
BALANCE		

NO. 541 $ 105.21
DATE *February 21,* 19 –
TO *National Telephone Co.*
FOR *January bill*

	DOLLARS	CENTS
BALANCE		
AMT. DEPOSITED		
TOTAL		
AMT. THIS CHECK		
BALANCE		

NO. 542 $ 211.97
DATE *February 23,* 19 –
TO *Eastern Power and Light Co.*
FOR *Electric bill*

	DOLLARS	CENTS
BALANCE		
AMT. DEPOSITED	659	28
	411	85
TOTAL		
AMT. THIS CHECK		
BALANCE		

NO. 543 $ 429.85
DATE *February 27,* 19 –
TO *Motors Acceptance Corp*
FOR *Truck payment*

	DOLLARS	CENTS
BALANCE		
AMT. DEPOSITED	896	51
TOTAL		
AMT. THIS CHECK		
BALANCE		

Business Applications

3. You spend one night each in Washington and Wheeling. Although you deposited the reimbursement for travel, it has not cleared your bank. How much is your account understated?

What business travel costs

Highest and lowest per diem for business travelers:

MOST EXPENSIVE

New York (Manhattan) $330
Washington $247
Honolulu $231

LEAST EXPENSIVE

Wheeling, W.Va. $76
Augusta, Ga. $76
Johnson City, Tenn. $80

Note: Per diem consists of breakfast, lunch, dinner and single-rate lodging in a business-class hotel.

Source: Runzheimer Meal-Lodging Cost Index of 200 U.S. cities

By Patti Stang and Marty Baumann, USA TODAY

Student Notes

SKILLBUILDER 6.2

Bank Statement Reconciliation

Learning Outcomes

After completing this skillbuilder, you should be able to:

■ Reconcile a bank statement

Reconciling Bank Statements

Every month the bank sends a statement for each checking account, summarizing all transactions that have taken place since the previous statement. The bank statement should promptly be compared to the checkbook record. Adjusting the checkbook balance and the bank statement balance is called **reconciliation.** The differences in the balances that may appear can be due to outstanding checks written but not yet cleared by the bank, deposits in transit (deposits made too late to be included on the statement), service charges, credits, or interest earned on the account.

MATH TIP

If there is an error in the checkbook balance, add or subtract the amount of the error at the end of the register rather than changing each amount.

A typical bank statement is shown below. The form printed on the back, which is used to assist in reconciling the account is shown on page 255. The bank statement shows a balance of $515.67. The balance on the last checkbook entry is $528.95. Use the following steps in reconciling the balances.

FIRST NATIONAL BANK OF SECAUCUS
526 Franklin Avenue
Secaucus, NJ 07094
(201) 555-1717

Reginald Perry
532 Sesame Lane
Secaucus, NJ 07094

ON	YOUR BALANCE WAS	NO.	WE SUBTRACTED CHECKS TOTALING	LESS ACTIVITY CHARGE	NO.	WE ADDED DEPOSITS OF	MAKING YOUR PRESENT BALANCE
Sep.10,19--	561.71	15	1220.77	0.00	4	1174.73	515.67

DATE	CHECKS				DEPOSITS	BALANCE
Sep. 10						561.71
Sep. 11	27.19					534.52
Sep. 16					318.53	853.05
Sep. 18	4.00	22.20				826.85
Sep. 21	15.00					811.85
Sep. 22	25.00				126.59(T)	913.44
Sep. 24	124.45	5.18(S)				783.81
Sep. 25	24.05					759.76
Sep. 30					225.00	984.76
Oct. 1	5.10	10.00	29.40		500.53	1440.79
Oct. 6	443.80					996.99
Oct. 7	18.99	100.00				878.00
Oct. 8	101.84	264.57				511.59
Oct. 10					4.08(I)	515.67

NAME: ——————————————————————— DATE: ———————————

SOLUTION

BANK RECONCILIATION FORM

Bank Statement Balance → 515.67 A

1. Write the bank statement balance at A on the reconciliation form.

Period ended

Add deposits in transit.

Date of deposit:	Amount.
10/10	250.00

2. Compare the bank statement and checkbook record. List any deposits not shown on the bank statement (**deposits in transit**), and total them at B. (A deposit of $250 was made late on October 10 and is not shown on the bank statement.)

Total 250.00 B
Total (A plus B) → 765.67 C

3. Determine the subtotal by adding the statement balance and deposits in transit (A + B). Write the subtotal at C.

List outstanding checks.

Check No.	Amount.
442	49.42
452	37.07
453	24.74

4. List any outstanding checks (three checks are outstanding, #442 for $49.42, #452 for $37.07, and #453 for $24.74 totaling $111.23), and enter the total at D.

Total 111.23 D

Current Bank Balance (C minus D) → 654.44

5. Deduct the total of outstanding checks (D) from the subtotal (C).

6. Total all credits and corrections on bank statement that increase the balance in the checkbook (automatic transfer—$126.59 and NOW interest $4.08). Add this total to the checkbook balance.

Checkbook	$528.95
⑥ Credit	+ 130.67
	659.62
⑦ Service charge	− 5.18
	$654.44

7. Total all charges and fees on the bank statement ($5.18 service charge), and subtract this amount from the checkbook balance.

8. The adjusted balances should now agree.

▶ **Self-Check**

The Happy Dog, a pet store, had a bank statement balance of $2,367.94. Deposits in transit totaled $1,619.88. Outstanding checks totaled $2,817.42. The account earned $10.58 in interest, and the checkbook balance was $1,149.82. Do the adjusted balances agree?

Self-Check Answer

no

Problems

Reconcile each checkbook balance with the bank balance.
Give the new checkbook balance.

Answers

1. Checkbook balance: $1,938.03   Bank balance: $1,796.43   **1.** _____
   Unreturned checks: $260.10; $400; $10.98; $18.50; $2
   Interest: $7.94
   Deposits not shown: $841.12

2. Checkbook balance: $218.08   Bank balance: $463.85
   Unreturned checks: $14.52; $126.78; $43.92; $165.80   **2.** _____
   Service Charge: $5
   Deposits not shown: $100.25

3. Checkbook balance: $632.95   Bank balance: $99.38
   Unreturned checks: $15; $7.21; $6; $18
   Service charge: $2
   Deposits not shown: $577.78   **3.** _____

Calculator Tip

To add or subtract money values, set your calculator on the Add/Mode position. If there is no Add/Mode, set the decimal selector at two.

4. Barry Goldman started a pet-walking service. At the end of the first month, his checkbook showed a balance of $146.92. When he received his bank statement, he found that he had been charged $18.75 for personalized checks, $5 service charge, and $10 for overdrawing his account during the month. What should his bank balance actually be?

Answers

4. _____

Business Applications

Complete the following bank reconciliations.

5. Karl's Lawnmower Service received a bank statement showing a balance of $605.50, from which a service charge of $7.50 had been deducted. Karl's Lawnmower Service's checkbook showed a balance of $448.16. Three checks were outstanding, totaling $164.84.

6. Dauber Dairies' bank statement showed a balance of $1,755.97. Two notes for $300 and $515 had been collected by the bank that month and credited to the account. There was a collection charge of $7.50 and a service charge of $4.50. The checkbook balance was $765.53. Outstanding checks totaled $187.44.

7. Julia's Catering Service's checkbook showed a balance of $224.03. The bank statement indicated a balance of $312.45, which included a service charge of $5.42. Two checks were outstanding totaling $309, and Julia's deposit of $215.16, dated April 25, was in transit.

8. The Corliss Corporation received its monthly bank statement showing a balance of $4,055.16. A $50 check, which had been deposited, was returned for insufficient funds. The bank also collected a $125 note for Corliss, for which they charged $3.50. The checkbook balance was $4,026.36, with outstanding checks totaling $157.30. There was a deposit in transit in the amount of $200 dated January 30.

BANK RECONCILIATION FORM

Bank Statement Balance → A

Period ended

Add deposits in transit.

Date of deposit: Amount.

Total   B

Total (A plus B) → C

List outstanding checks.

Check No.   Amount.

Total   D

Current Bank Balance (C minus D) →

BANK RECONCILIATION FORM

Bank Statement Balance → A

Period ended

Add deposits in transit.

Date of deposit: Amount.

Total   B

Total (A plus B) → C

List outstanding checks.

Check No.   Amount.

Total   D

Current Bank Balance (C minus D) →

BANK RECONCILIATION FORM

Bank Statement Balance → A

Period ended

Add deposits in transit.

Date of deposit: Amount.

Total   B

Total (A plus B) → C

List outstanding checks.

Check No.   Amount.

Total   D

Current Bank Balance (C minus D) →

BANK RECONCILIATION FORM

Bank Statement Balance → A

Period ended

Add deposits in transit.

Date of deposit: Amount.

Total   B

Total (A plus B) → C

List outstanding checks.

Check No.   Amount.

Total   D

Current Bank Balance (C minus D) →

Business Applications

9. If 90% of the people using PCs at home in 1996 have a software package that does bank reconciliations, what overall percent would have bank statement packages?

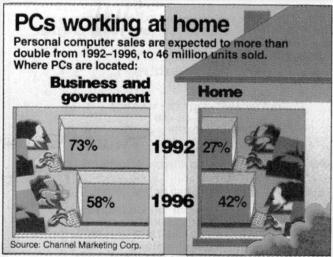

PCs working at home

Personal computer sales are expected to more than double from 1992–1996, to 46 million units sold. Where PCs are located:

	Business and government	Home
1992	73%	27%
1996	58%	42%

Source: Channel Marketing Corp.

By Deirdre Schwiesow and Sam Ward, USA TODAY

Student Notes

SKILLBUILDER 6.3

Payroll Plans

Learning Outcomes

After completing this skillbuilder, you should be able to:

- Compute gross earnings on an hourly wage basis.
- Compute gross overtime earnings.
- Compute gross earnings on a piecework basis.

Hourly Rate Payroll Plan

Many employees are paid a fixed hourly rate of pay, or a **regular hourly wage rate. Total (gross) earnings** are calculated by multiplying the hours worked by the wage for each hour worked.

EXAMPLE

Nannette Moreno worked the following hours one week: 8, 7, 8, 7, and 8. At a rate of $9.80 an hour, how much did she earn that week?

SOLUTION

8 + 7 + 8 + 7 + 8 = 38 h

$$\begin{array}{rl} \$9.80 & \text{Hourly rate} \\ \times \quad 38 & \text{Hours} \\ \hline \$372.40 & \text{Gross earnings} \end{array}$$

 **Self-Check**

Ken Sandovan worked $8, 7\frac{1}{2}, 8\frac{1}{2}, 8$, and 6 h one week. At a rate of $7.75 per hour, how much did he earn that week?

Self-Check Answer

$294.50

Problems

Complete the following partial payroll register to find each employee's **gross earnings.** The employee's actual take-home pay will be less than the gross earnings once deductions are made for social security, income taxes, and so on.

1.

Employee Number	Hours Worked					Regular Hours	Hourly Rate	Gross Earnings
	M	T	W	TH	F			
375	8	8	8	8	8	_____	$ 8.00	_____
411	8	8	8	8	4	_____	9.25	_____
465	8	$7\frac{1}{2}$	8	6	8	_____	11.50	_____
505	4	4	8	8	8	_____	8.45	_____
573	8	8	7	8	8	_____	7.25	_____
							Total	_____

Overtime

Employees receive their regular wage rate for time worked within the normal work-week, which is usually 40 hours. Time worked beyond the normal 40 hours is called **overtime.**

Federal law provides that employees shall be paid at a rate at least 50% greater than their regular hourly rate for all hours in excess of 40 per week. This is called **time and a half.** For example, at time and a half, a normal hourly rate of $9.80 would give an overtime rate of $14.70 an hour (9.80 × 1.5 = $14.70). When employees work on Sundays or holidays, they sometimes receive overtime rates that are double their regular hourly rate.

MATH TIP

You can write time and a half as 1.5 or as $1\frac{1}{2} = \frac{3}{2}$ to simplify computation.

EXAMPLE

During one week, Dee Morrison worked 45 h. If her regular hourly rate is $8.50, what were her gross earnings for the week if she is paid time and a half for all hours worked over 40?

SOLUTION

Her overtime rate is $8.50 × 1.5 = $ 12.75
Regular earnings   40 h × $8.50 = $340.00
Overtime earnings   5 h × $12.75 = $ 63.75
Total gross earnings $403.75

Problems

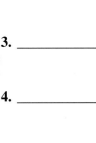

Self-Check

Roberto Sanchez worked $46\frac{1}{2}$ h one week. If his regular hourly rate is $6.88, what were his gross earnings for the week if he is paid time and a half for all hours over 40?

Self-Check Answer

_____$342.28_____

Find the salary in each case if overtime (hours over 40) is paid time and a half.

Answers

2. 46 h; $5.88 per hour

2. _____

3. 39 h; $9.55 per hour

3. _____

4. 42 h; $8.48 per hour

4. _____

5. 45 h; $4.98 per hour

5. _____

6. Andrea earns $4.22 per hour and double time if she works on Sunday. How much did she earn if she worked 39 h, 4 of which were worked on Sunday?

6. _____

Piecework Payroll Plan

Manufacturing employees are sometimes paid according to the number of items they complete. This is the **piecework wage plan.** The rate paid may vary with the difficulty of the task or the experience of the employee.

Problems

EXAMPLE

During a 5-day workweek, Marv Grasso completed 74 items. If the rate for each piece was $2.775, what were Marv's gross earnings?

SOLUTION

$74 \times \$2.775 = \$205.35$

 **Self-Check**

Betty Wilson completed 99 items and was paid $2.775 for each item. Find her gross earnings.

Self-Check Answer

$274.73

Calculator Tip

You can find overtime wages in a single operation by multiplying hours by 1.5 by the hourly wage. Then use the memory to store this value. Calculate and store regular wages and then use memory recall to find the total.

7. A partial payroll register for Singleton Instrument Company is shown below. Determine the total number of articles made by each employee. Determine gross earnings by multiplying total articles times rate. Round off gross earnings to the nearest cent.

SINGLETON INSTRUMENT COMPANY PAYROLL REGISTER

Employee Number	Number of Articles					Total Articles	Rate	Gross Earnings
	M	T	W	TH	F			
51	64	62	65	61	65	_____	$2.00	_____
52	64	65	60	62	—	_____	2.20	_____
53	68	65	66	65	62	_____	2.10	_____
54	59	62	60	64	66	_____	2.25	_____
55	62	65	63	59	56	_____	$2.455	_____
56	58	—	55	52	60	_____	2.50	_____
57	64	68	69	64	65	_____	2.25	_____
58	65	61	66	66	64	_____	2.775	_____
							Total	_____

Business Applications

8. The payroll register for Rob's Appliance Repair is shown here. Employees work overtime when the company has more repair projects to do than can be completed during a normal working day. Complete the payroll.

Employee Number	Hours Worked M T W TH F					Time Regular	Overtime	Hourly Rate	Wage Regular	Overtime	Gross Earnings
110	8	8	8	8	8	_____	_____	$8.50	_____	_____	_____
111	8	8	8	9	4	_____	_____	9.75	_____	_____	_____
112	8	9	9	8	8	_____	_____	11.00	_____	_____	_____
113	9	10	8	9	8	_____	_____	10.20	_____	_____	_____
114	3	8	9	9	10	_____	_____	14.00	_____	_____	_____
115	10	9	9	9	9	_____	_____	12.35	_____	_____	_____
116	8	8	8	8	8	_____	_____	7.50	_____	_____	_____
117	—	10	10	9	9	_____	_____	7.10	_____	_____	_____
118	8	9	$8\frac{1}{2}$	8	8	_____	_____	8.25	_____	_____	_____
119	9	9	9	9	7	_____	_____	15.00	_____	_____	_____
								Totals			

Business Applications

9. If the salespeople at Hewlett-Packard earn 2% of their sales, how much would they have earned in the fourth quarter of 1993?

Answer

9. _____

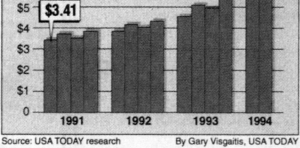

On a hot streak

After stumbling in the late 1980s, Hewlett-Packard has taken off the past three years.

Quarterly revenue in billions

$6.25

$3.41

Source: USA TODAY research   By Gary Visgaitis, USA TODAY

Student Notes

SKILLBUILDER 6.4

Payroll Deductions

Learning Outcomes

After completing this skillbuilder, you should be able to:

■ Use appropriate tables to compute payroll deductions.

Computing Payroll Deductions

A **payroll register** for Lund's Tool and Die is shown on page 268. There are four columns in the Employee Data section. The Number column is the number assigned to each employee by the employer. This number is commonly the individual's social security number. The remaining columns show the employee's name, marital status, and the number of exemptions claimed for tax withholding purposes.

The Earnings section contains four columns. The Hours column shows the total number of hours the employee worked during the pay period. The Regular column shows the employee's regular earnings—40 or fewer hours worked times the hourly rate. The Overtime column shows any overtime earned. The Total column shows the employee's total gross earnings—regular wages plus any overtime.

The Deductions section contains seven columns. FIT is the federal income tax deducted from the employee's total earnings. This deduction is determined from tables provided by the Internal Revenue Service (IRS) based on the payroll period, marital status, and the number of exemptions claimed. Both federal and state income taxes are mandatory taxes collected on all taxable earnings without limit. Examples of these tax tables are shown in the Appendix.

SS Tax (FICA, or social security) is another mandatory tax, collected at a flat rate, without reference to marital status or exemptions. At this time, SS Tax (social security tax) is collected at a rate of 6.2% on the first $57,600 of gross earnings. SS Med Tax (Medicare Tax) is collected on a similar basis at a rate of 1.45% on gross earnings up to $135,000. Most states and some cities have instituted a personal income tax, the exact basis being determined by the individual state. Union dues are commonly deducted from members' paychecks. AAP is a voluntary program whereby employees who choose to participate contribute 1% of their gross earnings to fund a local social program.

The Amount column in the Net Pay section shows the employee's actual take-home pay, which is total earnings minus total deductions. The employee paycheck number is recorded in the Ck. No. column.

MATH TIP

You can find the social security and medicare taxes by multiplying gross earnings by the rates of 6.2% and 1.45%, respectively.

EXAMPLE

Sam Purdy is married and claims three exemptions. His gross earnings for the week are $671. How much federal income tax, social security tax, and medicare tax are withheld from his weekly earnings?

SOLUTION

Select the income tax table for married persons located in the Appendix. In the left-hand column, locate the wage bracket that contains Sam's gross earning, $671 ("At least $670 but less than $680"). Move across the table to the column headed "3." The tax to be withheld is $63. Next to find the social security tax, multiply $671 × 6.2% (671 × 0.062). The social security tax is $41.60. To find the medicare tax multiply $671 × 1.45% (671 × 0.0145). The medicare tax is $9.73.

 ### Self-Check

LuAnn Riggin is married and claims four exemptions. Her gross earnings for the week are $582. Find her net pay after federal income tax, social security tax, and medicare tax are deducted.

Self-Check Answer

$484.48

Problems

Using the tables provided in the Appendix, compute deductions and net pay for each of the following problems.

	Gross Earnings	Exemp.	Marital Status	Federal Tax	SS Tax	Med Tax	Total	Net Pay
1.	$469.90	1	M					
2.	710.00	5	M					
3.	500.01	8	S					
4.	611.75	0	M					
5.	578.20	3	S					
6.	411.80	2	S					

Calculator Tip

Using the ⊞% key and the memory keys on your calculator will make it easier to calculate deductions and net pay.

7. Jim Allen's gross earnings last week were $489.99 He is single and claims two exemptions. How much (a) income tax, (b) social security tax, and (c) medicare tax would be deducted from his pay? (d) What is his net pay?

8. Helen Baque worked 43 hours last week. Her hourly rate is $12.50, with overtime paid at a rate of time-and-one-half. Helen is married and claims four exemptions. What is her net pay?

Answers

7. a. _____

   b. _____

   c. _____

   d. _____

8. _____

LUND TOOL AND DIE
PAYROLL REGISTER

For the Week Beginning April 19, 19 ___ and Ending April 23, 19 ___ Paid April 29, 19 ___

| EMPLOYEE DATA | | | | EARNINGS | | | | DEDUCTIONS | | | | | | | NET PAY | |
NO.	NAME	MARITAL STATUS	EXEMP.	HOURS	REGULAR	OVERTIME	TOTAL	FIT	SS TAX	MEDICARE TAX	STATE TAX	UNION DUES	AAP	TOTAL	AMOUNT	CK. NO.
111	Ahrend, M.	M	2	40	572 00	— 00						20 00				
112	Blase, G.	M	3	40	600 00	— 00						— 00				
117	Charlton, H.	S	0	42	544 00	40 80						20 00	— 00			
120	Dulong, R.	M	4	41	568 00	21 30						20 00				
121	Grish, V.	S	3	42	528 00	39 60						20 00				
124	Garth, M.	M	2	38	562 40	— 00						20 00	— 00			
130	Lennis, J.	M	4	44	496 00	74 40						20 00				
132	Mild, S.	S	0	40	576 00	— 00						— 00	— 00			
135	Milton, G.	M	3	41	548 00	20 55						20 00				
136	Pitcher, D.	S	2	37	558 70	— 00						20 00				
142	Popham, P.	S	1	40	564 00	— 00						20 00				
148	Quentin, J.	M	5	43	512 00	57 60						20 00				
149	Ridley, J.	M	1	40	560 00	— 00						— 00				
152	Scott, W.	M	3	42	520 00	39 00						— 00	— 00			
154	Willard, J.	S	2	44	500 00	75 00						20 00				
	Totals															

Business Applications

9. Complete the payroll register for Lund's Tool and Die, shown on page 268, using the tables provided in the Appendix. State tax is computed at 3% of gross earnings. For those employees who participate, 1% of their gross earnings is deducted for AAP. Check numbers begin with number 1411. Total all money columns.

Answer

10. If a single person has a payroll deduction for 20% of his or her estimated medical bills, how much would need to be deducted from his or her monthly paycheck?

10. _____

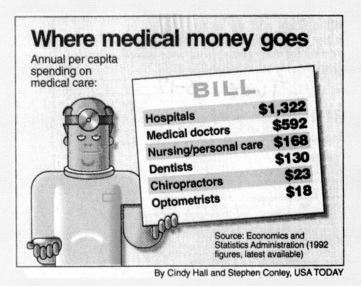

Where medical money goes

Annual per capita spending on medical care:

BILL

Hospitals	$1,322
Medical doctors	$592
Nursing/personal care	$168
Dentists	$130
Chiropractors	$23
Optometrists	$18

Source: Economics and Statistics Administration (1992 figures, latest available)

By Cindy Hall and Stephen Conley, USA TODAY

Student Notes

_____
_____
_____
_____
_____
_____
_____
_____
_____
_____
_____
_____

Student Notes

SKILLBUILDER 6.5

Taking Inventory

After completing this skillbuilder, you should be able to:

- Complete simple inventory records.
- Verify the accuracy of inventory records with a physical inventory.

Completing Inventory Records

A store's **inventory** is the quantity of goods or materials on hand. Taking inventory is the counting of goods or materials actually on hand. A **periodic inventory** requires counting the goods on hand at specific intervals, such as at year-end, semiannually, or quarterly. A **perpetual,** or running, **inventory** is one in which the counting is an ongoing process. In both cases, records, such as stock cards or computer runs, must be prepared and maintained. Computers and bar codes have made this task much easier, but not all businesses—especially smaller ones—lend themselves to a totally computerized operation. Knowing how much of a particular item is on hand is essential, and policies on when to reorder must be established in order to avoid being out of stock of an item.

MATH TIP

An incorrect calculation on inventory items can result in the loss of a sale or overstocking of items.

Baskets and More carries a large wicker basket that sells for $112.99. Complete the following inventory card for the basket.

Date	Quantity Received	Quantity Sold	Balance
5/4			2
5/7	16	13	
5/12	10	15	
5/19	25	18	

SOLUTION

On each day, add the quantity received to the previous balance and then subtract the quantity sold.

Date	Quantity Received	Quantity Sold	Balance
5/4			2
5/7	16	13	5
5/12	10	15	0
5/19	25	18	7

 **Self-Check**

On May 31, Baskets and More received 35 of the wicker baskets and sold 19. What was the inventory at the end of the day?

Self-Check Answer

__23__

Problems

Complete the four stock cards shown below. The minimum quantity is the reorder point. Circle any date on which any item falls below this amount.

1. Merchandise *Copy Paper, 5,000/case*

Minimum Quantity *36*

2. Merchandise *8½ × 11 Ruled Pads, 12/pack*

Minimum Quantity *140*

Merchandise *Copy Paper, 5,000/case*

Minimum *36*

Date	Quantity Received	Quantity Sold	Balance
19— 3/2			30
3/4	15		
3/8	18	10	
3/15	25	36	
3/19	20	28	
3/26	12		
3/29	36	24	

Merchandise *8½ × 11 Ruled Pads, 12 pack*

Minimum *140*

Date	Quantity Received	Quantity Sold	Balance
19— 3/3			155
3/9		110	
3/16	120		
3/18	120	90	
3/21		40	
3/25	180	50	
3/30		75	

3. Merchandise *#10 Envelopes; 50/pack*

Minimum Quantity *72*

4. Merchandise *Staples, 5,000/box*

Minimum Quantity *70*

Merchandise *#10 Envelopes, 50/pack*

Minimum *72*

Date	Quantity Received	Quantity Sold	Balance
19— 3/1			81
3/5		36	
3/9	40		
3/13	114	72	
3/22		50	
3/27	50	40	

Merchandise *Staples, 5,000/box*

Minimum *70*

Date	Quantity Received	Quantity Sold	Balance
19— 3/1			98
3/2	26		
3/11		48	
3/23	30	45	
3/24	59		
3/28		40	

Problems

5. Determine the total quantity in inventory for each item on the inventory sheet. Compare this to the minimum quantity indicated, and enter "Y" for "Yes" or "N" for "No" in the reorder column.

Item	Quan. at Loc. 1	Quan. at Loc. 2	Quan. at Warehouse	Total Quan.	Minimum Quan.	Reorder Y/N
Multifax #280388	8	2	7	_____	8	_____
Fax Modem #395731	4	6	5	_____	12	_____
Personal Copier #224402	1	5	3	_____	10	_____
Simpson Qt. Pr. #273052	2	1	0	_____	6	_____
Cordless Phone #284067	9	7	12	_____	30	_____
486SX/25 Comptr #322115	4	3	2	_____	10	_____

Physical Inventory

No matter what system is used for maintaining a running inventory, there are bound to be errors due to theft, breakage, mistakes in recording, and so on. To correct these errors, an actual count, or **physical inventory,** must be made of goods on hand. This is usually done at specified intervals but must be done whenever an accurate financial statement must be prepared.

Problems

Calculator Tip

Many businesses use computerized cash registers, laser readers, etc., to take inventory. However, these devices do not take stolen items into account.

EXAMPLE

Cramborski's Hardware lists 28 shovels in inventory that sell for $16.99. Is the total value of $473.72 listed on the inventory sheet correct?

SOLUTION

To find the total value, multiply the number in inventory by the unit costs:

$$\begin{array}{r} \$16.99 \\ \times\ \ 28 \\ \hline 135\ 92 \\ 339\ 8\ \ \\ \hline \$475.72 \end{array}$$

No, the total value on the inventory should be $475.72.

▶ **Self-Check**

Aaron's Art Store has 332 paintbrushes in stock that sell for $2.55. What value should be listed on the inventory sheet?

Self-Check Answer

$846.60

The two pages shown are part of the physical inventory taken at the end of the year. Check all extensions for accuracy, and enter any corrections in the space provided.

6.

PHYSICAL INVENTORY COUNT

SECTION Dinnerware
DATE Dec. 31, 19—

PART 12
SHEET NO. 3

COUNTED BY: F.H.   RECORDED BY: E.J.   COMPUTED BY: M.A.

STOCK NO.	QUANTITY	DESCRIPTION	UNIT PRICE	VALUE	CORRECTED TOTAL VALUE
24	20 sets	Lily	$ 27.88	$ 557.60	_____
33	23 sets	Stafford	62.00	1,326.20	_____
61	17 sets	Ironstone	89.00	1,513.00	_____
64	9 sets	Stoneware	54.40	504.27	_____
72	13 sets	Fuji	39.79	572.27	_____
80	39 sets	Pagoda	79.78	3,111.42	_____
93	11 sets	Bavaria	113.75	1,221.55	_____
		Total		$10,105.67	_____

Problems

7.

PHYSICAL INVENTORY COUNT

SECTION	Floor coverings				PART	15	
DATE	Dec. 31, 19—				SHEET NO.	5	CORRECTED TOTAL VALUE
COUNTED BY: M.R.		RECORDED BY: R.G.		COMPUTED BY: M.M.			

STOCK NO.	QUANTITY	DESCRIPTION	AREA OF EACH	UNIT PRICE	VALUE OF EACH UNIT	TOTAL VALUE	CORRECTED TOTAL VALUE
103	12	Carstar vinyl	99 ft^2	\$ 0.43/ft^2	\$ 42.57	\$ 510.84	_____
113	16	Bandon vinyl	99 ft^2	0.27/ft^2	26.73	427.68	_____
124	15	Econo vinyl	126 ft^2	2.17/yd^2	30.38	4,231.50	_____
162	9	Runners	42 ft^2	0.38/ft^2	15.96	146.44	_____
171	20	Manor tile, 12 ×12 in.	45 ft^2	0.39/ft^2	17.55	351.00	_____
178	19	Guild tile, 12 × 12 in.	25 ft^2	1.25/ft^2	31.25	593.75	_____
202	7	Drew tile, 12 × 12 in.	25 ft^2	0.95/ft^2	23.75	166.25	_____
204	43	Nylon shag	12 yd^2	16.49/yd^2	197.88	8,508.84	_____
255	27	Nylon broadloom	180 ft^2	17.93/yd^2	358.60	9,682.20	_____
		Total				\$24,618.50	_____

Business Applications

8. Complete the following inventory sheet for The Music Store order. Enter the quantity shipped and the balance remaining in inventory. If the quantity ordered is greater than the quantity in stock, ship what you have and back-order the difference.

Item	In Stock	Ordered	Shipped	Back Order	Balance
Port. Stereo RXFT510	38	2	_____	_____	
13-in. Color TV TC3866	8	13	_____	_____	
VCR VG2011	12	2	_____	_____	
CD Changer PD627	23	4	_____	_____	
Cassette Deck KWX3500	5	8	_____	_____	
Port. CD Player L8121	15	22	_____	_____	

9. Compute the value of each of the following physical inventory for Sagnaro Bed and Bath.

Item	Description	Unit Price	Stock Quantity	Actual Count	Value
M-3564	Mini blind	8.99	123	132	_____
RB3672	Roll-up blind	17.00	57	57	_____
V7884	Vertical blind	69.99	23	21	_____
PV-3260	Pleated shade	11.95	12	38	_____
BBT592	Comforter, twin	59.99	5	5	_____
BBQ692	Comforter, queen	79.99	4	2	_____
FL813T	Sheet, twin flat	5.99	27	27	_____
FF813T	Sheet, twin fitted	5.99	31	26	_____
JTS81	Juvenile twin set	21.99	15	15	_____
				Total	_____

10. Rachael's Tires compiled the following physical inventory sheet. Compute the value for each item based on the actual count.

Tire Item No.	Unit Price	Inventory Qty.	Actual	Value
P195/60HR14	$107.99	12	13	_____
P185/70SR13GT	53.99	21	21	_____
P215/75SR15	85.50	5	3	_____
33X12.50R15	115.95	2	2	_____
P165/80R13	33.75	35	39	_____
165SR13GT	42.00	7	8	_____
P195/70SR114BW	67.99	2	0	_____
LT235/85R16E	126.99	4	0	_____
P205/75R15XL	57.80	6	6	_____
			Total	_____

11. If you wanted to keep an inventory of 1,000 copies of *In the Kitchen With Rosie,* how many copies of *Cruel & Unusual* should you keep on hand?

Answers

11. _____

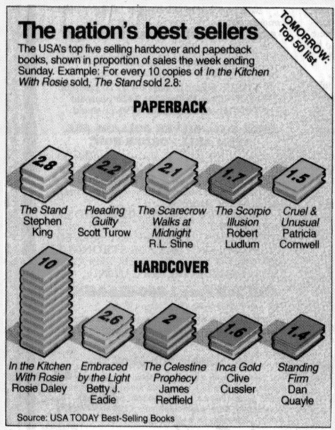

NAME: _____

DATE: _____

SKILLBUILDER 6.6
Valuing Inventory

Learning Outcomes

After completing this skillbuilder, you should be able to:

- Compute inventory value based on cost price or selling price.
- Compute inventory value based on an average (weighted average) basis.
- Compute inventory value based on a first in, first out basis.
- Compute inventory value based on a first in, last out basis.

Computing Inventory on Cost or Selling Price

When an inventory value is based on either cost price or selling price, the specific price of each lot of merchandise is used. The inventory value is found by multiplying the unit cost of each item by the number of items on hand. To find the total inventory value, add the inventory values of all items.

EXAMPLE

Compute the value of the following inventory based on the cost price.

SOLUTION

QUANTITY PURCHASED	ITEM	UNIT COST	ON HAND	INVENTORY VALUE
27 cases	Tomatoes	$16.56/case	11 cases	$182.16
36 cases	Pork and beans	11.76/case	21 cases	246.96
21 cases	Asparagus	21.36/case	9 cases	192.24
6 cases	Tuna	30.00/case	4 cases	120.00
			Total	$741.36

 **Self-Check**

The Big Barn carries three models of bicycles. They have 48 model A bikes in stock, selling for $118.75 each. They also have 27 of model B, selling for $199.99, and 11 of model C, selling for $244.99. Find the total inventory based on selling price.

Self-Check Answer

$13,794.62

Problems

Compute the inventory value of each item, and find the total inventory value for the following inventory sheets.

1.

INVENTORY REPORT AS OF JULY 1, 19—

Dept. Jewelry
Store No. 83

PURCHASE ORDER NO.	QUANTITY PURCHASED	ITEM	UNIT COST	ON HAND	INVENTORY VALUE
316	48	Watch, mens quartz	$65.00	12	_____
321	72	Watch, digital	26.00	39	_____
358	24	Watch, ladies quartz	90.00	9	_____
402	36	Fashion watch	18.11	15	_____
473	75	Juvenile watch	12.00	4	_____
				Total	_____

2.

INVENTORY REPORT AS OF JULY 1, 19—

Dept. Office Supplies
Store No. 83

PURCHASE ORDER NO.	QUANTITY PURCHASED	ITEM	UNIT COST	ON HAND	INVENTORY VALUE
516	8 doz.	Super Ball Pens	$21.46 doz.	3 doz.	_____
523	108 ea.	Halogen lamps	35.98 ea.	57	_____
524	10 doz.	Hi–Liters	16.20 doz.	5 doz.	_____
534	60 boxes	Computer paper	49.85 ea.	6	_____
547	84 boxes	DS/HD Disks/20	22.92 box	27	_____
				Total	_____

Computing Inventory Using the Average (Weighted Average) Method

When it is impractical to keep track of each purchase and its sale, the average, or weighted average, method is used. This method is used for a periodic inventory valuation and usually involves goods of relatively low value.

Problems

MATH TIP

When using average cost method, the average cost per unit must be rounded to the nearest cent.

EXAMPLE

What is the value of the inventory of paintbrushes as of June 30?

ITEM	Paintbrush		LOCATION			MAX.	1200
			BIN NO. 772			MIN.	300

DATE	QUANTITY IN	UNIT COST	REQ. NO.	QUANTITY OUT	BALANCE
19—					
5/15	350	$1.75			350
5/28			12	113	237
6/3	450	1.80			687
6/10	450	1.90			1,137
6/15			15	975	162
6/30	300	2.00			462

SOLUTION

First find the value of all units purchased at each price. Total the number of units and the inventory values.

$$\frac{\text{No. of}}{\text{Units}} \times \frac{\text{Unit}}{\text{Cost}} = \frac{\text{Inventory}}{\text{Value}}$$

$$
\begin{array}{rcrcr}
350 & \times & \$1.75 & = & \$612.50 \\
450 & \times & 1.80 & = & 810.00 \\
450 & \times & 1.90 & = & 855.00 \\
\underline{300} & \times & 2.00 & = & \underline{600.00} \\
1,550 & \times & & = & \$2,877.50
\end{array}
$$

Then divide the total inventory value by the total number of units to find the average unit cost of each unit. Round to three decimal places.

$$\frac{\text{Total Inventory}}{\text{Value}} \div \frac{\text{Total}}{\text{Units}} = \frac{\text{Average}}{\text{Unit Cost}}$$

$$\$2,877.50 \div 1,550 = \$1.8564516 \rightarrow \$1.856$$

To find the inventory value on June 30, multiply the number of units on hand by the average unit cost.

$$\frac{\text{Units}}{\text{on Hand}} \times \frac{\text{Average}}{\text{Unit Cost}} = \frac{\text{Inventory}}{\text{Value}}$$

$$462 \times \$1.856 = \$857.472 \rightarrow \$857.47$$

 **Self-Check**

What is the value on the inventory report on page 282 of the inventory of measuring cups as of August 30?

Self-Check Answer

$1,592.80

Problems

DATE	QUANTITY IN	UNIT COST	REQ. NO.	QUANTITY OUT	BALANCE
19—					
8/1	1,376	$0.88			1,376
8/10			2.46	810	566
8/15	1,500	$0.91			2,066
8/19			2.58	1,306	760
8/30	1,000	$0.93			1,760

3. Compute the value of the following inventory as of July 30, using the weighted average method.

3. _____

ITEM	Fertilizer			
DATE	QUANTITY IN	UNIT COST	QUANTITY OUT	BALANCE
19—				
3/30	360 bags	$ 7.20		360
4/5			315	45
4/13	300 bags	8.80		345
4/18			295	50
4/27	530 bags	8.00		580
5/5			400	180
5/25	160 bags	10.30		340
5/29			100	240
6/8	400 bags	9.20		640
6/17			250	390
7/6	250 bags	8.40		640
7/30			150	490

Computing Inventory Using The First-In, First-Out (FIFO) Method

The FIFO method is also used for a periodic valuation and assumes that the first-, or earliest-, purchased merchandise is the first sold or used in filling orders. The most recently purchased goods remain in inventory.

Problems

EXAMPLE

Three different lots of merchandise are purchased at different prices: 10 units at $7 each, 5 units at $9 each, and 8 units at $8 each. The ending inventory is 15 units. What is the value of the ending inventory on a FIFO basis?

SOLUTION

Total purchases were 23 units (10 + 5 + 8 = 23). The ending inventory is 15 units; therefore, 8 units were sold (23 − 15 = 8). The assumption is that the 8 units sold during this time were all from the first lot purchased, so we work back from the most recently purchased lots to establish the inventory value.

Number of Units		Unit Cost		Inventory Values	
8	×	$8.00	=	$64.00	
5	×	9.00	=	45.00	
2	×	7.00	=	14.00	(8 sold)
15				$123.00	

▶ **Self-Check**

Self-Check Answer

Compute the value of the ending inventory for the following using the FIFO method:

$805

Beginning inventory	40 units at $20
First purchase	50 units at $21
Second purchase	50 units at $22
Third purchase	50 units at $23
Ending inventory	35 units

Answer

4. Compute the value of the ending inventory for the following using the FIFO method.

4. _____

Beginning inventory	42 units at $12
First purchase	35 units at $18
Second purchase	15 units at $13
Third purchase	28 units at $15
Ending inventory	48 units

Computing Inventory Using the Last-In, First-Out (LIFO) Method

Another periodic method of valuing inventory is LIFO, which assumes that the last, or most recently purchased, merchandise is used to fill orders. The goods remaining in inventory would be from the lots purchased at the earliest dates.

Problems

Self-Check

Compute the value of the ending inventory for the following using the LIFO method.

First purchase	60 units at $44
Second purchase	30 units at $48
Third purchase	25 units at $42
Fourth purchase	40 units at $40
Ending inventory	72 units

Self-Check Answer

____ $3,216 ____

5. Compute the value of the ending inventory for the following using the LIFO method.

Beginning inventory	168 units at $30
First purchase	372 units at $24
Second purchase	218 units at $16
Third purchase	127 units at $42
Ending inventory	545 units

Answers

5. _____

Calculator Tip

When rounding to the nearest cent, set the decimal selector on the calculator on two places.

Business Applications

6. Cable Products' ending inventory of sound shields was 86 units. Beginning inventory was 35 units at $19.25 each. In May, 49 units were purchased at $18.75 each; in July, 60 units at $18.25 each; in October, 55 units at $19.75 each; in December, 62 units at $18.50 each. Compute the value of the ending inventory using the LIFO method.

Answers

6. _____

7. _____

8. _____

7. Cable Products' ending inventory of safety lanterns was 212 units. Beginning inventory in January was 102 units at $24 each. In April 84 units were purchased at $23 each; in June, 78 units at $24.50 each; in August, 108 units at $22 each; in November, 80 units at $22.50 each. Compute the value of the ending inventory using the FIFO method.

8. Complete the balance for each of the transaction dates, and then compute the average inventory value as of May 30.

ITEM	Long–Handle Shovel			
DATE	**QUANTITY IN**	**UNIT COST**	**QUANTITY OUT**	**BALANCE**
19—				
4/1	43	$ 6.19		43
4/7	154	6.19		_____
4/16			26	_____
4/23			41	_____
4/28	180	6.27		_____
5/5	40	6.25		_____
5/21			112	_____
5/30	100	7.11	56	_____

Business Applications

9. If each jar of jelly cost $0.75 to purchase and fill, complete the inventory value, based on cost, of 750,000 jars.

Berry popular jelly

$366.2 million

$291.5 million

Two berries account for more than half of the USA's yearly jelly purchases:

STRAWBERRY AND GRAPE

ALL OTHERS

Source: Nielsen Marketing Research

By John Riley and Bob Laird, USA TODAY

Student Notes

SKILLBUILDER 6.7

Computing Depreciation by Straight-Line and Double Declining Balance Methods

Learning Outcomes

After completing this skillbuilder, you should be able to:

- Compute depreciation by the straight-line method.
- Compute depreciation by the double declining balance method.

Depreciation

Depreciation is the loss incurred through the decline in value of property such as machinery, equipment, and buildings. It may also be due to obsolescence, which is the outdating of something because a new development or more modern item has replaced it. Four methods of computing depreciation are the straight-line method, the double declining balance method, the units-of-production method, and the sum-of-the-years' digits method.

Straight-Line Method

The **straight-line** method for computing depreciation includes the following steps.

1. Determine the probable, or **estimated, life** of the property.
2. Determine the **trade-in,** or **scrap, value** of the item. This is an estimate of the item's worth at the end of its useful life.
3. Find the difference between the original cost and the trade-in value. This difference is the total depreciation. Divide the total depreciation by the estimated (probable) life to find the **annual depreciation.**
4. Divide the annual depreciation by the total depreciation to find the **annual rate of depreciation.**

Problems

MATH TIP

Always remember to subtract the resale value from the original cost before dividing by the estimated life to calculate the annual depreciation.

1. Complete the following.

Original Cost	−	Trade-in Value	=	Total Depreciation
a. $ 23,850	−	$2,425	=	_____
b. 19,959	−	4,500	=	_____
c. _____	−	16,910	=	112,520
d. 51,859	−	_____	=	47,537

Problems

2.

Total Depreciation	÷	Estimated Life	=	Annual Depreciation
a. $ 6,516	÷	3 y	=	_____
b. 87,645	÷	15 y	=	_____
c. _____	÷	9 y	=	4,927
d. 112,288	÷	____ y	=	7,018

3.

Annual Depreciation	÷	Total Depreciation	=	Rate of Depreciation
a. $ 1,528	÷	$ 6,112	=	_____
b. 18,235	÷	455,875	=	_____
c. 348	÷	_____	=	8.33%
d. _____	÷	22,500	=	15%

4. Complete the following straight-line depreciation schedule.
Round the rate of depreciation to the nearest tenth of a percent
where necessary.

Property	Original Cost	Trade-in Value	Estimated Life	Annual Depreciation	Rate of Depreciation
Truck	$ 19,600	$1,960	9 y	_____	_____
Building	1,490,000	None	_____	_____	3.3%
Equipment	198,400	38,884	_____	13,293	_____
Car	19,000	8,360	4 y	_____	_____
Computer	_____	5,500	5 y	7,900	20%

Problems

Double Declining Balance Method

The double declining balance method is used to compute the annual depreciation on the decreasing value of an item. The trade-in value is not deducted for the cost of the item. This method involves the following steps.

1. Divide 100% by the estimated life of the item to determine the straight-line annual rate of depreciation. For example, a 5-year useful life is equivalent to 20% a year (100% ÷ 5 = 20%).
2. Double this rate (2 × 20% = 40%).
3. Multiply the original cost of the item by the double rate to find the first year's depreciation.
4. Subtract the first year's depreciation form the original cost to determine the value remaining at that time, which is the **book value** for the first year.
5. Repeat steps 3 and 4 for each successive year, beginning with the book value at the end of the first year. Continue this procedure until the book value equals the trade-in value.

EXAMPLE

A computer that costs $3,200 is expected to last 8 years. Use the double declining balance method to find the book value at the end of 3 years.

SOLUTION

1. 100% ÷ 8 = 12.5%
2. 12.5% × 2 = 25%, or 0.25

3. | *1st Year Depreciation* | = | *Cost* | × | *Double Rate* |
   |---|---|---|---|---|
   | $800 | = | $3,200 | × | 0.25 |

4. $3,200   Cost
   − 800   First-year depreciation
   $2,400   Book value at end of first year

5. $2,400 × 0.25 = $600   Second-year depreciation
   $2,400 − $600 = $1,800   Book value at end of second year
   $1,800 × 0.25 = $450   Third-year depreciation
   $1,800 − $450 = $1,350   Book value at end of third year

▶ **Self-Check**

A machine cost $12,000 and has an estimated life of 4 years. It is expected to have a trade-in value of $2,200 at that time. What is the book value of this machine at the end of the second year?

Self-Check Answer

$3,000

© by Glencoe/McGraw-Hill.

Problems

5. A piece of equipment cost $2,500 and is expected to last for 5 years. Using the double declining balance method, compute the depreciation for each of the 5 years, and show the book value at the end of the fifth year.

Year	Cost/Book Value	×	Rate of Depreciation	=	Depreciation for Year
	$2,500	×	_____	=	_____
	_____				
1. Bk. Val.	_____	×	_____	=	_____
	− _____				
2. Bk. Val.	_____	×	_____	=	_____
	− _____				
3. Bk. Val.	_____	×	_____	=	_____
	− _____				
4. Bk. Val.	_____	×	_____	=	_____
	− _____				
5. Bk. Val.	_____				

6. If the equipment in Problem 5 had a trade-in value of $500, (a) what would the amount of depreciation be at the end of year 4, and (b) what would the book value be at that time?

Answers

6. a. _____

b. _____

Problems

Calculator Tip

Most calculators allow mixed operations in a continuous process. After ⊕ or ⊖, press the total key. Then press the ⊗ or ⊘ key and enter the multiplier or divisor.

7. A lift truck originally cost $81,500 and has an estimated trade-in value of $7,500 after 10 years. Find the book value of the truck at the end of the third year.

Answers

7. _____

8. A retail check-out terminal costing $2,500 has a trade-in value of $450 at the end of 3 years. What is the book value at the end of the second year?

8. _____

9. A computer that originally cost $13,500 is estimated to have a trade-in value of $2,500 at the end of 6 years. Determine its book value at the end of the fourth year.

9. _____

Business Applications

Use the straight-line method of depreciation.

10. Pat's Floral Shop bought a station wagon costing $17,200. After 4 years, the car had a resale value of 44% of the original cost.
    a. Find the resale value.
    b. Find the annual depreciation.
    c. Find the rate of depreciation to the nearest tenth of a percent.

10. a. _____
    b. _____
    c. _____

11. The annual rate of depreciation on a warehouse was estimated to be 3.2% of the original cost. If the warehouse cost $2,850,000 what was the annual depreciation?

11. _____

12. A heavy-duty copy machine cost Loeb Sportswear $108,000 installed and is estimated to have a 15-y useful life, with a salvage value of $6,000. What are (a) the annual depreciation and (b) the rate of depreciation?

12. a. _____
    b. _____

Business Applications

13. A new, four-door hatch-back costs $17,250. If the car is depreciated over five years, using straight-line depreciation, how much is each year's depreciation?

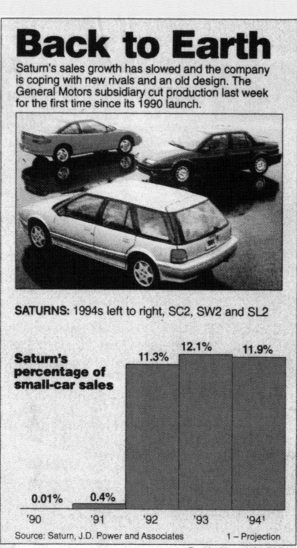

Back to Earth

Saturn's sales growth has slowed and the company is coping with new rivals and an old design. The General Motors subsidiary cut production last week for the first time since its 1990 launch.

SATURNS: 1994s left to right, SC2, SW2 and SL2

Saturn's percentage of small-car sales

0.01%	0.4%	11.3%	12.1%	11.9%
'90	'91	'92	'93	'94¹

Source: Saturn, J.D. Power and Associates   1 – Projection

By J.L. Albert, USA TODAY

Student Notes

_____

_____

_____

_____

_____

_____

_____

SKILLBUILDER 6.8

Computing Depreciation by Units of Production and Sum-of-the-Years' Digits Methods

Learning Outcomes

After completing this skillbuilder, you should be able to:

- Compute depreciation using units of production, either units produced or hours in operation.
- Compute depreciation using the sum-of-the-years' digits method.

Units of Production Method

The useful life of equipment is sometimes estimated in terms of the total amount of work performed either in the number of units produced or the number of hours of operation. This way of computing depreciation is known as the **units of production** method.

If a machine wears out after it has produced 50,000 units, its **production life** is 50,000 units.

The depreciation for the year based on units produced is determined by the following steps:

1. Determine total depreciation by subtracting the trade-in value from the cost.
2. Determine the unit depreciation by dividing total depreciation by the production life in units.
3. Find the depreciation for the year by multiplying the units produced by the unit depreciation rate.

Depreciation can also be based on hours of operation. To find depreciation based on hours of operation, follow these steps:

1. Find the total depreciation.
2. Divide total depreciation by production life in hours to find hourly depreciation.
3. Find the depreciation for the year by multiplying the hours used by the hourly depreciation.

Problems

EXAMPLE

a. If a machine costs $28,600, has a production life of 50,000 units, and has an estimated trade-in value of $1,400, determine the depreciation in a year in which it produces 4,750 units.

b. If the machine described in part (a) has a useful production life of 17,500 hours and was operated for 825 hours during one year, find the depreciation based on hours of operation.

SOLUTION

a. 1.
$$\begin{array}{ll} \$28,600 & \text{Cost} \\ -1,400 & \text{Trade-in value} \\ \hline \$27,200 & \text{Total depreciation} \end{array}$$

2.
$$\begin{array}{ccccc} \textit{Unit} & & \textit{Total} & & \textit{Production} \\ \textit{Depreciation} & = & \textit{Depreciation} & \div & \textit{Life in Units} \\ \$0.544 & = & \$27,200 & \div & 50,000 \end{array}$$

3.
$$\begin{array}{ccccc} \textit{Depreciation} & & \textit{Units} & & \textit{Unit} \\ \textit{For Year} & = & \textit{Produced} & \times & \textit{Depreciation Rate} \\ \$2,584 & = & 4,750 & \times & \$0.544 \end{array}$$

b. 1. Total depreciation = $27,200 [part (a)].

2.
$$\begin{array}{ccccc} \textit{Hourly} & & \textit{Total} & & \textit{Production} \\ \textit{Depreciation} & = & \textit{Depreciation} & \div & \textit{Life in Hours} \\ \$1.554285 \rightarrow \$1.55429 & = & \$27,200 & \div & 17,500 \end{array}$$

3.
$$\begin{array}{ccccc} \textit{Depreciation} & & \textit{Hours} & \times & \textit{Hourly} \\ \textit{for Year} & = & \textit{Used} & & \textit{Depreciation} \\ \$1,282.289 \rightarrow \$1,282.29 & = & 825 & \times & \$1.55429 \end{array}$$

▶ Self-Check

The production life of a machine costing $78,000 is estimated at 180,000 units. Its trade-in value is estimated at $6,000, and its estimated production life in hours is estimated at 42,000 h.

a. If yearly production is 4,440 units, find the depreciation.

b. If it is used 1,500 h in one year, find the depreciation.

Self-Check Answers

a. $1,776

b. $2,571.43

Problems

1. The production life of a machine is estimated at 180,000 units. The machine costs $78,000 and has an estimated trade-in value of $6,000. Complete the depreciation schedule below.

Year	Production	Depreciation
1	5,088 units	_____
2	3,510 units	_____

2. Compute the depreciation for the same machine if its estimated production life is 42,000 hours. It was used for the periods listed below. Round the hourly depreciation to 5 decimal places.

Year	Hours of Use	Depreciation
1	600	_____
2	2,200	_____

3. A machine that cost $17,700 has an estimated production life of 25,000 hours, with no trade-in value at the end of that time. If the machine was used 680 hours last year, find the depreciation for that year. Carry out the hourly depreciation to 3 decimal places.

Answers

3. _____

4. An automatic casting machine has a production life of 3,000,000 units. The machine cost $75,000 and has an estimated trade-in value of $1,500. Find the depreciation if 480,000 units were produced the first year. Carry out the unit depreciation rate to 4 decimal places.

4. _____

5. What is the book value at the end of the first year for the casting machine in Problem 4?

5. _____

Problems

The Sum-of-the-Years' Digits Method

The **sum-of-the-years' digits** method assumes the greatest depreciation for the years when an article is newest and its value is likely to drop the most. The rate of depreciation decreases each year. To compute depreciation using this method, follow these steps.

1. Estimate the useful life of the article in years.
2. Find the sum of all the digits of the years of estimated useful life. For example, for a machine with an estimated useful life of 4 years, the sum of the digits would be $1 + 2 + 3 + 4 = 10$.
3. Form a depreciation fraction by using the sum of the digits (in this case, 10) as the denominator of the fraction. The numerator is the digit of each of the individual years in reverse order: year $1 = \frac{4}{10}$, year $2 = \frac{3}{10}$, year $3 = \frac{2}{10}$, and year $4 = \frac{1}{10}$.
4. Subtract the estimated trade-in value from the original cost to find the total depreciation.
5. Multiply the total depreciation by the depreciation fraction to determine the depreciation for the first year.
6. Subtract the depreciation from the original cost to obtain the book value at the end of the first year.
7. Repeat steps 5 and 6, as necessary, to find the depreciation and book value for the required years.

MATH TIP

The sum of *n* years can be found by using the formula *n*(*n* + 1)/2. For 5 y, the sum is 5(5 + 1)/2 = 30/2 = 15.

EXAMPLE

A truck costs $34,800. After 5 years, it will have a trade-in value of $5,900. Construct a depreciation schedule using the sum-of-the-years' digits method.

SOLUTION

$1 + 2 + 3 + 4 + 5 = 15$

$34,800	Cost
− 5,900	Trade-in value
$28,900	Total depreciation

Year	Depreciation	Book Value at End of Year
1	5/15 × $28,900 = $9,633.33	$34,800.00 − $9,633.33 = $25,166.67
2	4/15 × 28,900 = 7,706.67	25,166.67 − 7,706.67 = 17,460.00
3	3/15 × 28,900 = 5,780.00	17,460.00 − 5,780.00 = 11,680.00
4	2/15 × 28,900 = 3,853.33	11,680.00 − 3,853.33 = 7,826.67
5	1/15 × 28,900 = 1,926.67	7,826.67 − 1,926.67 = 5,900.00*

*Trade-in value

Problems

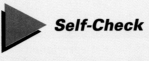

Self-Check

A delivery truck was purchased at a cost of $24,000. The salvage value after 4 y was expected to be $3,000. Compute the amount of depreciation for each year.

Self-Check Answers

Year 1 __$8,400__

Year 2 __$6,300__

Year 3 __$4,200__

Year 4 __$2,100__

6. If the useful life of an item is 8 y, find
   a. The sum-of-the-years' digit
   b. The fraction for the first year's depreciation.

Answers

6. a._____

b._____

7. A solar heating installation cost $30,500 and has an estimated life of 20 y, with a trade-in value of $2,500 at the end of that time. Find the depreciation for (a) the first year, (b) the second year, and (c) the third year.

7. a._____

b._____

c._____

8. An item that cost $1,250 has a trade-in value of $25 at the end of 7 y.
   a. Find the book value for the third year.
   b. Find the amount of depreciation for the sixth year.

8. a._____

b._____

Problems

Calculator Tip

To calculate with a fraction, write the fraction as a decimal with the calculator; press the total key, then do the next operation.

9. Complete the following depreciation table for an industrial grinder that cost $4,200 and will have a salvage value of $750 at the end of 4 y.

Year	Total Depreciation	×	Depreciation Rate	=	Annual Depreciation	Book Value
					0	$4,200
1	_____	×	_____	=	_____	_____
2	_____	×	_____	=	_____	_____
3	_____	×	_____	=	_____	_____
4	_____	×	_____	=	_____	_____

Business Applications

10. Coronet Mfg. purchased a machine that cost $86,800 and had a trade-in value of $6,500 after producing 550,000 units. Complete the following depreciation schedule.

Year	Units Produced	×	Depreciation per Unit	=	Depreciation for the Year	Cumulative Depreciation	Book Value
					0	0	$86,800
1	40,000	×	_____	=	_____	_____	_____
2	160,000	×	_____	=	_____	_____	_____
3	60,000	×	_____	=	_____	_____	_____
4	36,000	×	_____	=	_____	_____	_____
5	32,000	×	_____	=	_____	_____	_____

Business Applications

11. The State Department of Transportation bought a road grader that cost $165,000 and is expected to be operated for 50,000 h before is must be refurbished. At that time, its engine will be rebuilt at a cost of $42,000. What is (a) the cost per hour of rebuilding as compared to (b) the cost per hour to replace the machine if its replacement cost is still $165,000?

12. A courier vehicle owned by Roadrunner Express was driven 35,000 mi in its first year of operation. The depreciation charge for the year was $4,200. The expected useful life of the vehicle is 150,000 mi, with a trade-in value of $500. What was the original cost of the vehicle?

13. A $20,000 car is bought and depreciated over 5 years. Compute the book value after 2 years of the sum-of-the-digits method.

Answers

11. a. _____

 b. _____

12. _____

13. _____

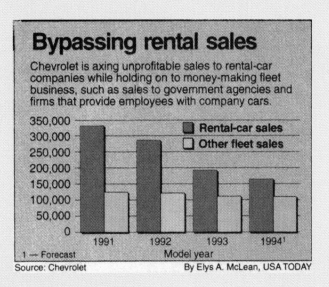

Bypassing rental sales

Chevrolet is axing unprofitable sales to rental-car companies while holding on to money-making fleet business, such as sales to government agencies and firms that provide employees with company cars.

■ Rental-car sales
□ Other fleet sales

Model year

1 — Forecast

Source: Chevrolet   By Elys A. McLean, USA TODAY

Student Notes

_____

_____

_____

_____

_____

U N I T

7

INTEREST AND DISCOUNTS

In this unit we discuss how to compute simple and compound interest using a variety of strategies, as well as how to discount noninterest-bearing and interest-bearing notes. Finally, we learn how to compute cash and trade discounts.

According to the figure shown here, 41% percent of Americans responding to a survey believe that rising interest rates affect them negatively. What are some reasons for this? How are you affected by rising interest rates?

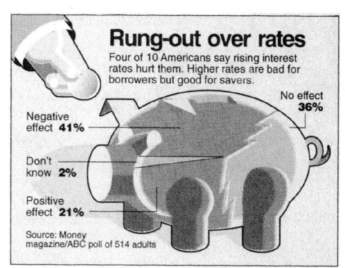

Rung-out over rates
Four of 10 Americans say rising interest rates hurt them. Higher rates are bad for borrowers but good for savers.

No effect **36%**

Negative effect **41%**

Don't know **2%**

Positive effect **21%**

Source: Money magazine/ABC poll of 514 adults

By Marty Baumann, USA TODAY

Bakery Owner

A former baker, Kelly, decides to open her own bakery. As an entrepreneur, Kelly must be able to market products, bake bread and pastries, serve customers, and take care of the business' finances. Kelly will need to project earnings and cash required to start the business and supplement earnings until a profit is earned.

$$\frac{\text{1st month interest} \times (\text{Number of payments} + 1)}{2}$$

Math Application

Kelly decides that she would like to expand her business, but that delivery of the bakery items would be necessary. Kelly decides to price some vans and trucks. Kelly finds a used van priced at $11,000. She can finance the van for 6% interest over 3 years. If Kelly makes a 15% down payment, what will be her monthly payments?

First, compute the loan amount.

Price times 15% = down payment

Now compute the loan amount.

Purchase price – down payment = loan amount

Compute the interest and monthly payment.

1st month interest = Principal × Rate × Time

Interest on unpaid balances.

Monthly Payment = (Principle + Interest)/Number of months toward loan.

Critical Thinking Problem

Kelly can lease a new van for $250 a month for 36 months, or purchase a last year's van for $300 a month for 60 months. What are some issues that Kelly should consider before making a final decision?

SKILLBUILDER 7.1

Computing Simple Interest by Formula

After completing this skillbuilder, you should be able to:

- Compute interest using the interest formula and either a 12-month year or a banker's year.
- Find the exact time between two dates.
- Compute interest using the ordinary interest method.

Simple Interest and the Banker's Year

Interest can be considered the rent paid to borrow money. The **principal** is the amount of money borrowed, and the **rate** is the percent of interest charged.

Time is the period of time for which the money is on loan. Since many loans are for less than one year, time may be expressed as a fraction of the year in terms of days, weeks, or months. Use this formula to compute interest.

$$Interest = Principal \times Rate \times Time$$

or

$$I = P \times R \times T$$

When the time of the loan is expressed in days, the number of days of the loan is divided by the number of days in a year. For purposes of computing interest, a **banker's year** or **commercial year** of 360 is used.

Problems

MATH TIP

When computing interest, the interest rate is written as a decimal.

EXAMPLE

Find the interest on $1,269 at 9% for 5 mo.

SOLUTION

$$I \quad = \quad P \quad \times \quad R \quad \times \quad T$$

$$\$47.59 \ = \ \$1,269 \ \times \ 0.09 \ \times \ \frac{5}{12}$$

EXAMPLE

Find the interest on $3,200 at 11% for 90 days, using the banker's year.

SOLUTION

$$I \ = \ P \quad \times \quad R \quad \times \quad T$$

$$\$88 \ = \ \$3,200 \ \times \ 0.11 \ \times \ \frac{90}{360}$$

 **Self-Check**

a. Find the interest on $8,000 at 6% for 8 mo.
b. Find the interest on $4,600 at 8.5% for 180 d, using the banker's year.

Self-Check Answers

a. $320

b. $195.50

Compute the interest in these problems using the forumla $I = P \times R \times T$. Use the banker's year (360 d) when days are indicated. Round percents to the nearest tenth of a percent, where necessary.

1. $1,850 at 9% for 90 d

2. $600 at 11% for 75 d

3. $3,500 at 6% for 145 d

4. $7,500 at 12% for 5 mo

5. $11,000 at $7\frac{3}{4}$% for 212 d

6. $3,853 at 15% for 1 y 4 mo

Answers

1. _____
2. _____
3. _____
4. _____
5. _____
6. _____

Problems

Answers

7. $5,780 at $10\frac{1}{4}$% for 182 d   **8.** $25,167 at 8% for 47 d

9. $1,927 at $6\frac{3}{8}$% for 135 d   **10.** $970 at $10\frac{1}{2}$% for 30 d

11. $11,680 at $9\frac{1}{4}$% for 11 mo   **12.** $7,827 at 13% for 7 mo

7. _____

8. _____

9. _____

10. _____

11. _____

12. _____

Finding the Exact Time

The exact time is found by counting the actual number of days between any two dates, excluding the first date but including the last date.

When counting days, remember that the month of February has 29 d instead of 28 during a leap year. A leap year occurs every 4 yr. If the number of a year is evenly divisible by 4, it is a leap year. (There is an exception. If the year ends in at least two zeros, for example, 1800, 1900, or 2000, it must be divisible by 400 to be a leap year.) For example,

$$1996 \div 4 = 499 \qquad \text{1996 is a leap year}$$

EXAMPLE

Find the exact number of days between December 12, 1995, and March 28, 1996.

SOLUTION

From December 12   19 d left in December
                             31 d in January
                             29 d in February (leap year)
To March 28   28 d in March
                         107 d, exact time

 **Self-Check**

Find the exact number of days between January 5, 1995, and March 3, 1995.

Self-Check Answer

57

Problems

Find the exact number of days between the following dates.

13. May 5 to September 17   **14.** October 23 to January 8

15. July 7 to November 29   **16.** January 4 to March 24

17. June 15 to August 23.   **18.** September 1 to December 29

13. _____

14. _____

15. _____

16. _____

17. _____

18. _____

Compute the interest due on the following loans using the banker's year and exact time.

19. $2,400 at $7\frac{1}{2}$% from January 12 to May 27

19. _____

20. $600 at 11% from August 14 to December 1

20. _____

21. $11,750 at $9\frac{3}{4}$% from April 3 to May 27

21. _____

22. $1,241 at 14% from December 21 to March 18

22. _____

23. $32,975 at $7\frac{7}{8}$% from April 5 to July 23

23. _____

24. $1,490 at 6% from January 4 to February 15

24. _____

25. $9,875 at $11\frac{1}{2}$% from June 21 to November 5

25. _____

26. $28,653 at $10\frac{5}{8}$% from February 21 to May 5

26. _____

27. $2,050 at $8\frac{3}{4}$% from October 15 to January 5

27. _____

28. $17,962 at $9\frac{1}{8}$% from November 3 to March 3

28. _____

Problems

29. $8,125 at 13% from February 12 to August 31

30. $4,900 at $6\frac{7}{8}$% from July 9 to September 19

Answers

29. _____

30. _____

Ordinary Interest

Ordinary interest is frequently used in business because it is easier to count the time and determine the due date of loans. Ordinary interest assumes that a year consists of 12 mo of 30 d each. To find the date on which a loan is due, count forward the number of months of the loan. The loan will be due on the same day of the month as the loan began.

If the month the loan is due does not contain the same last day number as the month when the loan was made, then the last day of the month is used.

EXAMPLE

Find the due date of a 3-mo loan made on April 15.

SOLUTION

April 15 to May 15	1 month
May 15 to June 15	1 month
June 15 to July 15	1 month
	3 months

The loan is due on July 15.

EXAMPLE

Find the due date of a 2-mo loan made on July 31.

SOLUTION

July 31 to August 31	1 month
August 31 to September 30	1 month
	2 months

The loan is due on September 30.

The method used to count time may cause a difference of a day or two in the length of the loan, which will affect the amount of interest earned. For example, the 2-mo loan in this example was made on July 31 and was due on September 30. By ordinary interest, the time would be counted as follows: 2 mo × 30 d = 60 d. By the exact time method, the exact count of days from July 31 to September 30 would be 61 d (31 d in August plus 30 d in September).

 **Self-Check**

Find the due date of a 4-mo loan made on May 31.

Self-Check Answer

September 30

Problems

Determine the date on which each of the following loans will be due.

31. A 2-mo loan made on July 31

32. A 3-mo loan made on November 30

33. A 7-mo loan made on March 15

34. An 11-mo loan made on April 21

31. _____

32. _____

33. _____

34. _____

Using the ordinary interest method, determine (a) when the following loans are due and (b) the amount of interest to be paid.

35. $1,650 at 8% for 2 mo from April 15.

35a. _____

b. _____

36. $4,269 at 11% for 6 mo from August 31, 1995.

36a. _____

b. _____

37. $12,775 at $9\frac{1}{2}$% for 3 months from June 4

37a. _____

b. _____

38. $21,520 at $7\frac{1}{8}$% for 4 months from July 31

38a. _____

b. _____

39. $38,995 at $10\frac{3}{4}$% for 11 months from November 30

39a. _____

b. _____

40. $6,150 at $8\frac{5}{8}$% for 5 months from January 18

40a. _____

b. _____

Calculator Tip

Set the decimal-point selector at two when computing the interest amount.

41. $15,385 at $12\frac{1}{4}$% for 9 months from August 23

41a. _____

b. _____

42. $9,775 at $6\frac{7}{8}$% for 8 months from March 17

42a. _____

b. _____

Business Applications

43. McGill Consultants borrowed $12,500 to purchase new office furniture. The bank quoted an interest rate of $9\frac{1}{2}\%$ for 180 d. Using the banker's year, compute (a) the amount of interest McGill will be charged and (b) the total amount to be repaid.

43. a. _____

b. _____

44. On January 30 Knutsen's Insurance borrowed $2,800 for 1 mo at a rate of 9%. (a) Is the company paying more or less for this loan on an ordinary interest basis, as compared to paying interest for the exact number of days of the loan? (b) How much more or less will it pay?

44. a. _____

b. _____

45. Bernadette's Audiology applied for a commercial loan in the amount of $42,800 for a period of 2 y. Because it was an unsecured loan, she was quoted a rate of $10\frac{1}{2}\%$. If the loan was secured by pledging company assets, the rate would be reduced by 2 percentage points. How much would each loan rate cost her?

45. a. _____

_____

Business Applications

46. If $500 is invested at 7.01% simple interest, how much interest is earned over 6 months?

Answers

46. _____

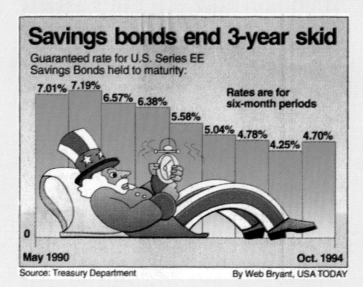

Savings bonds end 3-year skid

Guaranteed rate for U.S. Series EE
Savings Bonds held to maturity:

7.01% 7.19% 6.57% 6.38% 5.58% 5.04% 4.78% 4.25% 4.70%

Rates are for
six-month periods

0

May 1990   Oct. 1994

Source: Treasury Department   By Web Bryant, USA TODAY

Student Notes

SKILLBUILDER 7.2

Using a Simple Interest Table

Learning Outcomes

After completing this skillbuilder, you should be able to:

- Compute simple interest using a formula.
- Find the principal, rate, or time as the unknown.

Finding Simple Interest Using a Table

Simple interest can be computed by using a table such as the one shown in the Appendix on page 400. This table shows the interest earned for $100 at various rates of interest and for different periods of time. To use the simple interest table, follow these steps.

1. Find the appropriate Rate-of-Interest column at the top of the table.
2. Find the appropriate number of days or months of interest in the Time column.
3. Locate the point at which the Rate-of-Interest column and the Time column meet. This point indicates the amount of interest earned by $100 at that rate for the period of time.
4. Divide the principal by 100 (move the decimal point two places to the left) to determine the number of $100 units contained in the principal.
5. Multiply the interest on $100 by the factor obtained in step 4 to determine the amount of interest earned.

To determine the interest factor for time periods or interest rates not shown on the table, factors may be combined. The simplest combinations should always by used when combining factors.

MATH TIP

Remember, to determine interest factors for time and rates not shown in the table, combine factors. The simplest combinations should always be used.

EXAMPLE

Find the interest earned on $1,729.85 for 17 d at a rate of $6\frac{1}{2}$%. Round the answer to the nearest cent.

SOLUTION

Select the $6\frac{1}{2}$% column, and then move down to row 17. The factor of 0.3069

is the interest earned on $100 for 17 d at $6\frac{1}{2}$ percent.

$$\begin{array}{ccc} & & \textit{Interest} \\ \textit{Interest} & = & \textit{Loan} \times \textit{Factor} \\ \$5.3089096 \rightarrow \$5.31 & = & \$17.2985 \times 0.3069 \end{array}$$

Computing the same interest by the interest formula method gives the following result.

$$\begin{array}{ccc} I & = & P \times R \times T \\ \$5.309678 \rightarrow \$5.31 & = & \$1,729.85 \times 0.065 \times \dfrac{17}{360} \end{array}$$

This demonstrates that the table gives a sufficiently accurate result.

EXAMPLE

Using the simple interest table, find the interest earned on $3,000 at $10\frac{1}{2}$% for 72 days.

SOLUTION

$$10\frac{1}{2}\% = 4\% \qquad + \qquad 6\frac{1}{2}\%$$

72 days = 2 months (60 days) + 12 days

2 months =	60 days at 4%	= 0.6667
	60 days at $6\frac{1}{2}$%	= 1.0833
	12 days at 4%	= 0.1333
	12 days at $6\frac{1}{2}$%	= 0.2167
	72 days at $10\frac{1}{2}$%	= 2.1000

Interest = Loan divided by 100 × Interest Factor

$$\$63.00 = \$30.00 \qquad \times \qquad 2.100$$

Check: $3,000 × 0.105 × $\dfrac{72}{360}$ = $63.00

 **Self-Check**

Using the simple interest table, find the interest earned on $2,500 at 10% for 50 d.

Self-Check Answer

$34.72

Problems

Use the simple interest table on page 400 to compute the interest charges on the following loans. You will have to determine the exact number of days for Problems 13 through 16.

1. $2,500 at 11% for 48 d

    1. _____

2. $1,725 at 7% for 180 d

    2. _____

3. $8,000 at 14%, for 86 d

    3. _____

4. $426 at $10\frac{1}{2}$% for 75 d

    4. _____

5. $800 at 8% for 112 d

    5. _____

6. $2,980 at $9\frac{1}{2}$% for 99 d

    6. _____

7. $15,265 at $8\frac{1}{2}$% for 205 d

    7. _____

8. $7,458 at 12% for 141 d

8. _____

9. $1,850 at 10% for 54 d

9. _____

10. $4,200 at $12\frac{1}{2}$% for 32 d

10. _____

11. $11,875 at 13% for 93 d

11. _____

12. $1,183 at $9\frac{1}{2}$% for 67 d

12. _____

13. $3,850 at 11% from October 19 to April 17

13. _____

14. $504 at $13\frac{1}{2}$% from November 20 to March 2

14. _____

15. $8,250 at $11\frac{1}{2}$% from April 2 to June 13

15. _____

Problems

16. $402 at 9% from May 20 to September 9

Answers

16. _____

Finding the Principal, Rate, or Time

The following chart can be used to find the formula for the principal, rate, or time.

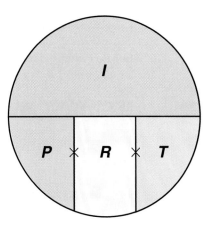

To find the **principal,** cover the letter *P*. The remaining letters indicate that the principal is found by dividing the interest by the product of the rate multiplied by the time.

$$P = \frac{I}{R \times T}$$

To find the **rate,** cover the letter *R* on the chart. The remaining letters indicate that the rate is found by dividing the interest by the product of the principal multiplied by the time.

$$R = \frac{I}{P \times T}$$

To find the **time,** cover the letter *T* on the chart. The remaining letters indicate that the time is found by dividing the interest by the product of the principal multiplied by the rate.

$$T = \frac{I}{P \times R}$$

The resulting number represents a fraction of a year and must be multiplied by 360 or 365 to determine the number of days for which the interest was paid.

Problems

EXAMPLE

If $42 interest was earned at a rate of 8% for 54 d, what was the amount of the loan?

SOLUTION

The amount of the loan, or the principal, is the unknown. Change 8% to a decimal fraction, and substitute the known quantities in the formula.

$$P = \frac{I}{R \times T}$$

$$= \frac{\$42}{0.08 \times \dfrac{54}{360}} = \frac{\$42}{0.012}$$

$$= \$3,500$$

EXAMPLE

For how many days was a loan outstanding if $1,750 earned $31.50 interest at a rate of 9%?

SOLUTION

$$T = \frac{I}{P \times R}$$

$$= \frac{\$31.50}{\$1,750 \times 0.09} = \frac{\$31.50}{\$157.50}$$

$$= 0.2 \text{ year or } 0.2 \times 360 \text{ days}$$

$$= 72 \text{ days}$$

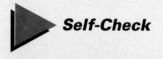 **Self-Check**

Self-Check Answer

_____7.5%_____

If $26.25 interest was earned on $3,000 for 42 d, what was the rate of interest?

SKILLBUILDER 7.3
Computing Exact Interest

Learning Outcomes

After completing this skillbuilder, you should be able to:

- Compute exact interest.
- Use a table to compute exact time.

Computing Exact Interest

The United States government always uses a 365-d year to compute interest. This is called **exact interest,** and it is more accurate than interest computed on a banker's year, since it is based on the true number of days in a year. To compute exact interest, use the simple interest formula $I = P \times R \times T$, but change the denominator of the time fraction from 360 to 365.

EXAMPLE
Compute the exact interest due on a loan of $3,500 at 12% for 103 d.

SOLUTION

$$I = P \times R \times T$$
$$\$118.520 \to \$118.52 = \$3,500 \times 0.12 \times \frac{103}{365}$$

 **Self-Check**

Compute the exact interest due on a loan of $4,100 at 9% for 225 d.

Self-Check Answer

$227.47

Problems

Compute the exact interest due on the following loans. Where necessary, count the precise number of days between dates.

Answers

1. $11,000 for 180 d at 12%

1. _____

2. $942 from July 10 to September 16 at $8\frac{1}{2}\%$

2. _____

3. $9,867 from August 20 to December 7 at $11\frac{1}{2}\%$

3. _____

4. $1,550 from November 14 to April 14 at 9%

4. _____

5. $15,000 from May 4 to August 10 at 8%

5. _____

6. $6,000 for 240 d at $8\frac{1}{2}\%$

6. _____

7. $1,286 for 72 d at 10%

7. _____

8. $775 for 48 d at $7\frac{1}{2}\%$

8. _____

Problems

Using exact time and exact interest, find the missing element in each of the following problems.

	Principal	×	Rate	×	Time	=	Interest
9.	$7,000	×	_____	×	90 d	=	$120.82
10.	$4,000	×	9%	×	_____	=	313.64
11.	_____	×	7%	×	68 d	=	20.87
12.	$12,800	×	$11\frac{1}{2}$%	×	_____	=	350.86

Using a Table to Compute Exact Time

The table on page 401 can be used to count the exact number of days between two dates. Each day of the year is consecutively numbered, from 1 for January 1 to 365 for December 31. To use the table, follow these steps.

1. Select the column with the appropriate month heading.
2. Move down that column to the row that is the date of the day you are seeking. (Note that the extreme left and right columns are numbered consecutively 1 to 31.)
3. The intersection of the date row and month column is the consecutive number of that day in the year.
4. For the second date, repeat steps 1 to 3.
5. Subtract the smaller number from the larger. The difference is the exact number of days between the two dates.

The table assumes a 365-day year. If the loan period includes the month of February in a leap year, add one to the consecutive number of each day after February 28. February 29 would be day 60, March 1 would become day 61, and so on.

Problems

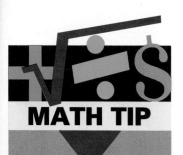

MATH TIP

Remember, the table assumes a 365-d year. If the loan period includes February in a leap year, add a day to the total number of days.

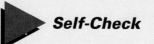

EXAMPLE

What is the exact number of days between November 28 and March 15?

SOLUTION

Go to the column headed "November," and move down to row 28. November 28 is day 332. Subtract 332 from 365 to count 33 days in the old year. March 15 is day 74.

$$74 + 33 = 107$$

The number of days is 107. (If this were a leap year, March 15 would be day 75, and the total would then be 108 days.)

▶ **Self-Check**

Find the exact number of days between March 12 and December 12.

Self-Check Answer

_____275_____

Determine the exact number of days between the following dates.

Answers

13. March 12 and August 23   13. _____

14. September 8 and December 19   14. _____

15. January 8 and May 19 (leap year)   15. _____

16. December 27 and February 11   16. _____

17. April 15 and October 21   17. _____

Problems

18. November 25 and March 5

18. ——————————

Determine the exact time and interest for the following loans,
using the table on page 401 to count the days.

Amount of Loan	Rate	Term of Loan	Method	Number of Days	Interest Due
19. $ 3,590.00	8%	8/12 to 11/9	Banker's	———————	———————
20. 1,527.00	$7\frac{1}{2}\%$	1/18 to 4/11 (nonleap year)	Exact	———————	———————
21. 450.00	11%	12/14 to 2/28	Exact	———————	———————
22. 5,872.00	10%	6/23 to 12/15	Banker's	———————	———————
23. 1,624.00	9%	4/26 to 6/12	Banker's	———————	———————
24. 8,527.00	$7\frac{3}{4}\%$	1/16 to 4/10	Exact	———————	———————
25. 37,693.00	$10\frac{1}{4}\%$	10/9 to 2/6	Banker's	———————	———————

Problems

Use exact time and exact interest unless otherwise indicated.

26. A $16,000 loan at 10% was made on May 13 and was repaid on July 12. How much interest was paid?

26. _____

27. A $5,000 loan was made on January 12. On April 12 a total amount of $5,093.75 was repaid. What rate of interest was charged?

27. _____

28. A $23,900 loan at $7\frac{3}{4}$% was charged $304.48 interest. For how long was the loan outstanding?

28. _____

Calculator Tip

On many calculators, it is not necessary to press the ⌷=⌷ key as you work. Enter the principal and press ⌷×⌷. Enter the rate and press ⌷%⌷ ⌷×⌷. Enter the number of days, press ⌷÷⌷, and enter 365. Then press ⌷=⌷.

29. $419.18 interest was paid on a 90-d loan at $8\frac{1}{2}$%. What was the amount of the loan to the nearest whole dollar?

29. _____

30. A loan of $86,427 was made for 69 d at 11% interest. How much interest would be paid (a) using the banker's year method and (b) using the exact interest method?

30. a. _____

b. _____

Business Applications

31. The Last National Bank of Seattle loaned $375,000 to one of its best corporate customers. Interest was charged at one and a half points over prime for a period of 32 d. Prime rate is 6%. Using the exact interest basis, how much interest would they pay?

31. _____

32. Michelle Carlson, treasurer of Juliet Jewelry, has discovered an error in Juliet's tax payments. The additional $13,000 due is subject to a penalty of 25% of the amount owed, plus exact interest from April 15 to July 21, the date of the payment. What is the total amount that must be paid?

32. _____

33. Shaydee Enterprises' tax return for the previous year was audited by the IRS. As a result of this audit on October 19, it was determined that they owed an additional $28,000 in tax. In addition to the tax payment, penalty interest is charged at 120% percent of the normal rate of 9% from April 15 to the audit date plus a further penalty of 5% per month, not to exceed 25%. What was the total amount of the payment due from them?

33. _____

34. If your money is being invested, do you prefer calculating interest using the banker's year or exact interest?

Answer

34. _____

Who owns stocks?

Although there's been an explosion of mutual funds lately, funds own only 10% of stocks:

Individuals 54%

Pension funds 25%

10% Mutual funds

5% Foreign investors

4% Insurance companies

2% Other

Source: USA TODAY research

By Sam Ward, USA TODAY

Student Notes

SKILLBUILDER 7.4

Computing Compound Interest

Learning Outcomes

After completing this skillbuilder, you should be able to:

- Compute compound interest using a formula.
- Compute compound interest using a periodic rate.
- Compute compound interest using a table.
- Find interest compounded daily using a table.

Using a Formula

When money is deposited in a savings bank or other savings institution, the account earns **compound interest.** This means that the interest made on an account is periodically combined with the principal so that the base amount on which interest is paid becomes greater and greater. To compute compound interest, follow these steps.

1. Find the simple interest on the principal for one interest period, using the simple interest formula $I = P \times R \times T$.
2. Add this interest to the first period principal to find the principal for the second period.
3. Compute the interest earned on the second period principal.
4. Add the second period interest to the second period principal to find the third period principal.
5. Repeat this procedure for each period for which compound interest is to be earned. The total of the original principal and all interest earned is the **compound amount.** To find the compound interest earned, subtract the original principal from the compound amount.

When interest is compounded annually, the value of T is 1. However, interest is usually compounded more than once a year. When this happens, replace the time factor, T, in the simple interest formula by a common fraction representing the compounding period (semiannual, $\frac{1}{2}$; quarterly, $\frac{1}{4}$; monthly $\frac{1}{12}$; daily, $\frac{1}{365}$).

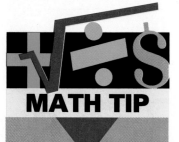

MATH TIP

When compounding interest, the interest must be added to the principal at the end of each interest period.

EXAMPLE

A deposit of $5,000 is made to a savings account earning interest at 4% compounded annually. How much would be in the account at the end of 2 y if the interest is left in the account? What is the compound interest?

SOLUTION

1. *Year 1:* $5,000 × 0.04 × 1 = $200 interest
2.   + 200   First year's interest
3. *Year 2:* $5,200 × 0.04 × 1 = $208 interest
4.   + 208   Second year's interest
              $5,408   Compound amount
5. $5,408   Compound amount
            − 5,000   Original principal
              $408   Compound interest

Simple interest would have resulted in only $400 interest ($200 for each year).

► **Self-Check**

Find the interest if $15,000 is compounded monthly at 5% for 1 y.

Self-Check Answer

$767.41

Problems

Find (a) the compound amount and (b) the compound interest for these accounts. Round all money amounts to the nearest cent where necessary.

Answers

1. $1,500 at 4% compounded annually for 3 y

1. a. _____

   b. _____

2. $6,000 at 3% compounded annually for 4 y

2. a. _____

   b. _____

3. $3,000 at $2\frac{1}{2}$% compounded semiannually for 4 y

3. a. _____

   b. _____

Problems

4. $12,000 at 6% compounded quarterly for 2 y

4. a. _____

   b. _____

5. $4,500 at 3% compounded bimonthly for 2 y

5. a. _____

   b. _____

Problems

6. $8,000 at 4% compounded monthly for 1 y

6. a. _____

    b. _____

7. $11,000 at $3\frac{1}{2}\%$ compounded every 4 months for 2 y

7. a. _____

    b. _____

Problems

Using a Periodic Rate

Compound interest can be computed by using the rate of interest for a compounding period in the formula instead of the annual rate.

EXAMPLE

A deposit of $5,000 is made to a savings account earning interest at 4% compounded semiannually for 1 y. Find the compound amount and the compound interest.

SOLUTION

1. Determine the number of compounding periods in 1 y (semiannually = 2 periods a year).
2. Divide the annual rate of interest by the number of compounding periods in a year (4% ÷ 2 = 2%).
3. Add 1 to the period rate found in step 2 (0.02 + 1 = 1.02). When the principal on deposit is multiplied by a factor such as this, the product is the compound amount—principal plus interest—for that period.
4. Multiply the total time on deposit by the number of compounding periods in 1 y to determine the number of compounding computations to be made (1 y × 2 periods a year for semiannual compounding = 2 compounding periods).
5. Multiply the beginning principal by the factor obtained in step 3. Continue to multiply the resulting product by this factor for the required number of compounding periods obtained in step 4.

Period 1: $5,000 × 1.02 = $5,100
Period 2: $5,100 × 1.02 = $5,202   Compound amount
                      − 5,000   Original principal
                    $ 202   Compound interest

 **Self-Check**

Find (a) the compound amount and (b) the compound interest if $5,000 earns 4% interest and is compounded semiannually for 2 y.

Self-Check Answers

a. $5,412.16
b. $412.16

Problems

Using the preceding method, find (a) the compound amount and (b) the compound interest for these accounts. Carry out the periodic rate to four decimal places where necessary.

Answers

8. $850 at 3% compounded semiannually for 2 y

8. a. _____

b. _____

9. $3,600 at 5% compounded every 3 months for 2 y

9. a. _____

b. _____

10. $7,500 at 6% compounded quarterly for $\frac{3}{4}$ y

10. a. _____

b. _____

Problems

11. $9,875 at $2\frac{1}{2}\%$ compounded bimonthly (every 2 months) for $\frac{2}{3}$ y

11. a. _____

b. _____

12. $1,500 at $4\frac{1}{2}\%$ compounded monthly for 8 mo

12. a. _____

b. _____

13. $12,000 at 3% compounded monthly for $\frac{1}{2}$ y

13. a. _____

b. _____

14. $1,850 at 4% compounded every 4 months for 1 y

14. a. _____

b. _____

Problems

Using the Compound Interest Table

The table on page 402 lists factors needed to compute compound interest at various rates for up to 25 interest-compounding periods.

EXAMPLE

Compute the compound amount and compound interest on $5,580 at 6% compounded quarterly for 5 y.

SOLUTION

1. Determine the number of compounding periods in 1 y (quarterly = 4).
2. Divide the annual rate of interest by the number of compounding periods in 1 y to get the interest rate per period (6% ÷ 4 = $1\frac{1}{2}$%).
3. Multiply the total number of years on deposit by the number of compounding periods in 1 y to get the total number of compounding periods (5 y × 4 = 20).
4. Locate the appropriate interest rate column in the table on page 402 ($1\frac{1}{2}$%). Move down the rate column to the line matching the total number of periods determined in step 3, in this case, 20. The factor at this point is the value of $1 compounded at that rate for that period of time: 1.346854.
5. Multiply the original principal by the compound factor to get the compound amount.

Compound Amount = Original Principal × Compound Factor

$7,515.45	=	$5,580	×	1.346854

$7,515.45   Compound amount
− 5,580.00   Original principal
$1,935.45   Compound interest

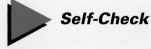 **Self-Check**

Determine (a) the compound interest and (b) the compound amount on $4,000 at 6% compounded bimonthly for 2 y.

Self-Check Answers
a. ___$507.30___
b. ___$4,507.30___

Problems

Use the table on page 402 to find the compound amount and the compound interest for the following accounts.

Principal	Annual Int. Rate	Time on Deposit	Compounded	Compound Amount	Compound Interest
15. $ 8,000	4%	4 y	Quarterly	_____	_____
16. 4,100	2%	12 y	Semiannual	_____	_____
17. 1,260	5%	6 y	Quarterly	_____	_____
18. 6,975	3%	$12\frac{1}{2}$ y	Semiannual	_____	_____
19. 7,500	2%	3 y	Quarterly	_____	_____
20. 12,000	$2\frac{1}{2}$%	2 y	Semiannual	_____	_____
21. 4,500	4%	2 y	Monthly	_____	_____
22. 5,000	$4\frac{1}{2}$%	3 y	Every 2 mo	_____	_____
23. 3,700	2%	1 y 8 mo	Every 4 mo	_____	_____
24. 4,000	4%	1 y	Monthly	_____	_____
25. 4,000	3%	$3\frac{1}{2}$ y	Every 2 mo	_____	_____
26. 38,500	$4\frac{1}{2}$%	3 y	Every 4 mo	_____	_____

Problems

Finding Interest Compounded Daily

On page 403 is a compound interest table for daily compounding of interest. It is customary to credit the interest earned at the end of each month when interest is compounded daily. To use this table, select the appropriate annual rate column: then read down the column to the number of days for the compounding factor.

EXAMPLE

How much interest would $5,000 earn at a rate of 3% compounded daily for 30 d?

SOLUTION

Read down the 3% column to line 30. The factor is 1.0024659. Then multiply the principal by the compounding factor to find the compound amount.

$5,000 × 1.0024659 = $5,012.33 Compound amount

Subtract the original principal from the compound amount to find the compound interest.

$5,012.33   Compound amount
− 5,000.00   Original principal
   $12.33   Compound interest

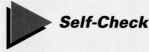 **Self-Check**

How much would $2,500 earn at a rate of 3.5% compounded daily for 25 d?

Self-Check Answer

$5.99

Problems

Use the daily compounding table on page 402 to find the compound amount and the compound interest earned by these accounts. Use seven decimal places for the compound factor.

Calculator Tip

Set the round-off switch on the calculator at four decimal places.

Principal	Annual Int. Rate	Time on Deposit	Compound Amount	Compound Interest
27. $ 8,000	$2\frac{1}{2}\%$	23 d	_____	_____
28. 11,500	3%	31 d	_____	_____
29. 35,000	$3\frac{1}{2}\%$	5 d	_____	_____
30. 1,300	2%	29 d	_____	_____
31. 19,800	$2\frac{3}{4}\%$	19 d	_____	_____
32. 14,159	$3\frac{1}{2}\%$	12 d	_____	_____

Business Applications

33. Weber Industries placed $15,000 in a savings account that paid 4% compounded monthly for 1 y in order to finance equipment purchases at the end of the year. Find the compound interest and the compound amount in the account at the end of the year.

33. _____

_____

_____

34. In order to plan for an anticipated expansion, Lori's Boutique put $22,000 in an account in which interest was paid at 3% compounded quarterly. Find the compound interest and the compound amount at the end of 9 mo.

34. _____

_____

_____

Business Applications

35. A family of four invests $1,000 in an account that earns 8.5% interest, compounded monthly. How many days could the family spend on vacation using the money earned from the compound amount one year later?

Answer

35. _____

On the road

Daily cost for lodging, meals and driving expenses for a family of four on vacation:

$215.30

'80   '85   '90   '94¹

Source: American   1—Projected
Automobile Association

By Suzy Parker, USA TODAY

Student Notes

SKILLBUILDER 7.5

Discounting Noninterest-Bearing Notes and Drafts

Learning Outcomes

After completing this skillbuilder, you should be able to:

- Find the proceeds of a noninterest-bearing note.
- Find the proceeds of a noninterest-bearing draft.

Finding the Proceeds of a Noninterest-Bearing Note

A **promissory note** is a written promise to pay a sum of money at a specific time. The date on which the payment is promised is the **due date,** or **maturity date.** The maturity date of the note may be specified exactly, or it may be stated as a definite number of days, months, or years from the date of issue of the note. The time period from the date of issue to the maturity date is the **term of the note.**

To find the maturity date of a 60-day note dated July 20, count the exact number of days from the date of issue to determine the date on which the note is to be paid. If the term of the note is stated in months, count forward from the date of the note the given number of months to the same date in the month of maturity. For example, a 2-mo note issued July 15 would be due September 15. A 2-mo note issued July 31 would be due September 30, since there are only 30 d in September. If cash is needed before the maturity date of the note, the business may sell the note to a bank. This process is called **discounting the note,** and the date of sale to the bank is the **discount date.** The **term of discount** is the exact number of days from the discount date to the maturity date.

When a note is discounted, the bank charges banker's interest, the **amount of discount.** The interest is computed on the face value for the term of discount and is then deducted from the face value of the note. The balance, called the **proceeds** of the note, is paid to the borrower. A typical noninterest-bearing note is shown on the next page.

Problems

$ 1,500.00 _____ April 17, 19 ___

___ Sixty (60) days ___ after date _J_ promise to pay to

the order of _Kris Koshak_

One thousand five hundred and 00/100 _____ Dollars

Payable at _Morgan State Bank_

Value received _with no interest_

No. _1785_ Due _June 16, 19___   _John A. Bailey_

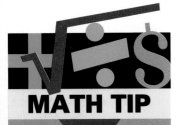

MATH TIP

When a note is discounted, use 360 d as the denominator.

EXAMPLE

Find the maturity date and term of discount of a 90-d note dated June 20 and discounted July 9.

SOLUTION

Use the table on page 401 to determine the exact number of days between two months.

June 20 to September 20 = 92 d
Less 2 d = September 18   **Maturity date**
July 9 to September 9 = 62 d
September 9 to September 18 = 9 d
62 d + 9 d = 71 d   **Term of discount**

If the same note ran for 3 mo and was discounted on the same date, the maturity date would be September 20 and the term of discount from July 9 to September 20 would be 73 d.

EXAMPLE

If the note signed by John Bailey was discounted on May 1 at 9%, find the proceeds.

SOLUTION

$1,500.00   Face value of note
              Maturity date: 60 d from April 17 = June 16
              Term of discount: May 1 to June 16 = 46 d

$- 17.25$ Less amount of discount: $1,500 \times 0.09 \times \dfrac{46}{360}$

$\overline{\$1,482.75}$ Proceeds

Problems

1. Refer to the promissory note on page 344. From this note, complete the following information:

a. Principal _____   **b.** Term of note _____
c. Date of Issue _____   **d.** Payee _____
e. Rate _____   **f.** Maker _____
g. Maturity date _____

2. Find the maturity date and term of discount for each of the following.

Date of Note	Term of Note	Maturity Date	Discount Date	Term of Discount
a. September 30	30 d	_____	October 15	_____
b. January 18,1996	45 d	_____	January 31	_____
c. March 31	3 mo	_____	May 4	_____
d. September 25	120 d	_____	October 25	_____
e. May 12	72 d	_____	June 15	_____
f. November 26	210 d	_____	March 26	_____
g. June 4	60 d	_____	July 5	_____
h. February 29	90 d	_____	April 1	_____
i. August 31	6 mo	_____	December 30	_____
j. October 1	54 d	_____	October 31	_____

Problems

3. Find the maturity date, term of discount, amount of discount, and proceeds for the following noninterest-bearing notes. Use the banker's year.

	Face Value	Term of Note	Date of Note	Discount Date	Rate of Discount	Maturity Date	Term of Discount	Amount of Discount	Proceeds
a.	$500	30 d	Jan. 5	Jan.15	9%	_____	_____	_____	_____
b.	1,750	90 d	Mar. 26	May 2	11%	_____	_____	_____	_____
c.	5,000	2 mo	Nov. 18	Dec.31	$7\frac{1}{2}\%$	_____	_____	_____	_____
d.	8,659	72 d	June 12	July 1	10%	_____	_____	_____	_____
e.	2,500	45 d	Sept. 8	Sept. 30	$6\frac{3}{4}\%$	_____	_____	_____	_____
f.	985	60 d	Oct. 15	Nov. 1	8%	_____	_____	_____	_____
g.	7,850	210 d	May 20	Aug. 26	$11\frac{1}{4}\%$	_____	_____	_____	_____
h.	12,500	4 mo	Aug. 31	Nov. 1	$9\frac{1}{2}\%$	_____	_____	_____	_____
i.	6,250	180 d	Feb. 12 1996	July 10	$10\frac{5}{8}\%$	_____	_____	_____	_____

Problems

Finding the Proceeds of a Noninterest-Bearing Draft

A **draft** is a written order for payment made by one party to another. It differs from a check in that it must be accepted before it becomes legally binding. If the draft calls for payment at a specified date, it is a **time draft,** in which case the date of acceptance is not important. When the draft calls for payment a certain time after sight, or acceptance, it is a **sight draft,** and the maturity date can be determined only after the draft has been presented for acceptance. The parties to a draft are the **drawer,** the party creating the draft; the **payee,** the party to whom payment is to be made; and the **drawee,** the party being ordered to pay. With a sight draft, the drawee is also the **acceptor.**

$ *1,500.00*	*April 17,* 19 ——
Sixty (60) days from date	Pay to the
order of *Kris Koshak*	
One thousand five hundred and ⁰⁰/₁₀₀ ————————	Dollars
Value received and charge to the account of	
To *Last National Bank of Secaucus*	
No. *782*	/s/ *John A. Bailey*

A draft is discounted in the same way as a promissory note.

Problems

EXAMPLE

Find the proceeds on a sight draft for $3,850. Its maturity date is 45 d after sight. It was accepted on November 15 and discounted on December 15 at 12%.

SOLUTION

$3,850.00 Face value of draft
Maturity date: 45 d from November 15 = December 30
Term of discount: December 15 to December 30 = 15 d

$$- 19.25 \quad \text{Less amount of discount: } \$3,850 \times 0.12 \times \frac{15}{360}$$

$3,830.75 Proceeds

▶ **Self-Check**

Find (a) the maturity date, (b) the term of discount, (c) the amount of discount, and (d) the proceeds for a $9,000 noninterest-bearing draft if the date of the draft is June 4, the due date is 4 mo from the date, the acceptance date is July 12, the discount date is July 12, and the rate of discount is $10\frac{1}{4}$%. Use the banker's year.

Self-Check Answers	
a.	Oct 4
b.	84 d
c.	$215.25
d.	$8,784.75

4. From the information provided on the time draft shown on page 347, identify the following:

a. Principal _____ **b.** Term of draft_____

c. Date of issue _____ **d.** Drawer _____

e. Payee _____ **f.** Drawee _____

g. Maturity date_____

Calculator Tip

You can enter percents in decimal form or use the whole-number percent and the percent key.

Problems

5. Find the maturity date, term of discount, amount of discount, and proceeds for the following noninterest-bearing drafts. Use the banker's year. Use the space on page 350 to work out your answers.

	Face Value	Date of Draft	Due Date	Acceptance Date	Discount Date	Rate of Discount	Maturity Date	Term of Discount	Amount of Discount	Proceeds
a.	$ 2,000	March 30	60 d from date	—	April 30	$8\frac{1}{2}\%$				
b.	5,500	September 12	45 d after sight	October 2	October 15	10%				
c.	1,200	February 15	3 mo from date	—	March 1	11%				
d.	875	October 9	90 d after sight	November 1	November 2	9%				
e.	3,750	May 17	72 d from date	—	June 1	$7\frac{1}{4}\%$				
f.	11,200	December 11	15 d after acceptance	February 8	February 10	$10\frac{3}{4}\%$				
g.	7,683	August 26	45 d from date	—	September 12	8%				
h.	1,150	April 5	10 d after sight	June 3	June 3	$6\frac{3}{4}\%$				
i.	2,900	November 28	30 d from date	—	December 1	$9\frac{1}{2}\%$				

Problems

Use the space below to solve Problem 5.

Business Applications

6. On May 24, Melinda Phelps was having cash-flow problems and was unable to pay the $7,800 she owed for the latest shipment of merchandise purchased from High Seas Imports. Since she is an excellent customer of High Seas Imports, they have agreed to extend credit to her for 72 d at no interest. However, Melinda must sign a promissory note for the credit extension. On maturity, the promissory note is to be paid into High Seas' account at the Last National Bank of Secaucus. On June 28, High Seas discounted the note at 9%. From the information provided, complete the following promissory note, and compute the proceeds.

$ _____   _____ 19 _____

_____ after date _____ promise to pay to

the order of _____

_____ Dollars

Payable at _____

Value received _____

No. _____ Due _____ 19 _____   _____

7. On September 11, the buyer for High Seas Imports arranged the purchase of $112,000 of merchandise from Zodiac Products. A 120-d time draft was given to Zodiac in payment for the shipment through High Seas' account at Buck's Bank International. On November 11, Zodiac discounted the draft with its bank at $10\frac{1}{4}\%$ interest. From the information provided, complete the following draft and compute the proceeds that Zodiac received from its bank.

$ _____   _____ 19 _____

_____ Pay to the

order of _____

_____ Dollars

*Value received and charge
to the account of*

To _____ /

No. _____   _____ /   _____

8. You are self-employed and one of your suppliers took a non-interest bearing note from you. Is the lack of interest good or bad for your taxes?

8. _____

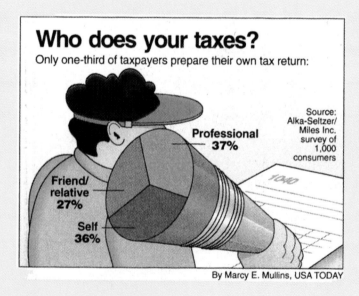

Who does your taxes?
Only one-third of taxpayers prepare their own tax return:

Professional 37%

Friend/relative 27%

Self 36%

Source: Alka-Seltzer/Miles Inc. survey of 1,000 consumers

By Marcy E. Mullins, USA TODAY

SKILLBUILDER 7.6

Discounting Interest-Bearing Notes and Drafts

Learning Outcomes

After completing this skillbuilder, you should be able to:

■ Find the proceeds on interest-bearing notes and drafts.

Finding the Proceeds on Interest-Bearing Notes

Promissory notes and drafts may or may not bear interest. When a borrower uses any of these as **collateral,** that is, security for a loan, the bank pays the proceeds to the borrower and collects the maturity value when the obligation is due. The **maturity value** of an interest-bearing note or draft is the face value plus interest earned for the period of the note or draft. The bank lends the maturity value less the amount of discount (interest) computed on the maturity value. The borrower receives the remainder, or **proceeds.**

When a note or draft is collected by a bank, some banks deduct a certain percent of the maturity value of the note as a **collection fee.**

Problems

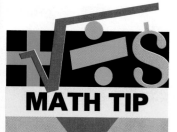

MATH TIP

When calculating the number of days from the discount date, look carefully at how time is stated; it may be stated in months or in exact days.

EXAMPLE

A $2,000, 90-d promissory note, dated July 2, with interest at 9%, is discounted on August 15 at $10\frac{1}{2}$%. Find the proceeds.

SOLUTION

Maturity date: 90 d from July 2 = September 30

$2,000.00	Face value
+ 45.00	Plus interest: $2,000 × 0.09 × $\frac{90}{360}$
$2,045.00	Maturity value
	Term of discount: August 15 to September 30 = 46 days
− 27.44	Less amount of discount: $2,045 × 0.105 × $\frac{46}{360}$
$2,017.56	Proceeds

EXAMPLE

A 30-d, $9\frac{1}{2}$% note with a face amount of $2,000 is discounted after 10 d at a rate of $10\frac{3}{4}$%. The bank charges a collection fee of 0.1% (0.001). Find the proceeds.

SOLUTION

$2,000.00	Face value
+ 15.83	Plus interest: $2,000 × 0.095 × $\frac{30}{360}$
$2,015.83	Maturity value
− 2.02	Less collection fee: $2,015.83 × 0.001
$2,013.81	
− 12.04	Less discount: $2,015.83 × 0.1075 × $\frac{20}{360}$
$2,001.77	Proceeds

 **Self-Check**

A $5,784, 120-d promissory note, dated October 31, with interest at 11%, is discounted on February 1 at $12\frac{1}{2}$%. Find the proceeds.

Self-Check Answer

$5,941.89

Problems

1. Find the proceeds for each of the following notes. Use the banker's year.

	a.	b.	c.	d.
Face value	$1,000	$1,920	$516	$296.40
Term of note	2 mo	60 d	3 mo	90 d
Rate of interest	$8\frac{1}{2}\%$	$10\frac{1}{2}\%$	9%	$7\frac{1}{4}\%$
Amount of interest	$ _____	$ _____	$ _____	$ _____
Maturity value	$ _____	$ _____	$ _____	$ _____
Date of note	November 5	June 19	May 30	February 21, 1996
Discount date	December 6	July 9	July 16	April 4
Rate of discount	10%	$11\frac{1}{2}\%$	$10\frac{3}{4}\%$	9%
Maturity date	_____	_____	_____	_____
Term of discount	_____ d	_____ d	_____ d	_____ d
Amount of discount	$ _____	$ _____	$ _____	$ _____
Proceeds	$ _____	$ _____	$ _____	$ _____

Problems

2. Lisa Hahn took a 120-d note for $1,500 from Sidney Wilde. The note was dated May 3, with interest at 9%. On May 18, Lisa Hahn sold the note to First Federal Bank at a discount rate of $10\frac{1}{4}\%$. What were the proceeds on the note?

Answers

2. _____

3. Find the proceeds for each of these drafts, using the banker's year.

	a.	b.	c.	d.	e.
Face value	$2,250	$3,840	$6,000	$1,769	$2,750
Date of draft	May 5	July 12	August 8	October 19	January 17
When due	75 d from date	2 mo after sight	3 mo from date	45 d after sight	180 d from date
Acceptance date	May 12	July 21	August 12	November 1	March 17
Rate of interest	$6\frac{3}{4}\%$	0%	$7\frac{1}{4}\%$	0%	9%
Amount of interest	$_____	$_____	$_____	$_____	$_____
Maturity value	$_____	$_____	$_____	$_____	$_____
Maturity date	_____	_____	_____	_____	_____
Discount date	May 15	August 4	August 15	November 8	May 15
Rate of discount	9%	$7\frac{3}{4}\%$	10%	$8\frac{1}{2}\%$	11%
Term of discount	_____ d	_____ d	_____ d	_____ d	_____ d
Amount of discount	$_____	$_____	$_____	$_____	$_____
Proceeds	$_____	$_____	$_____	$_____	$_____

Problems

4. Melody Manufacturers was given a 120-d, $8\frac{1}{2}$% note for $2,315. The note was dated July 14 and was discounted on September 1 at a rate of 11%. Find the proceeds.

Answers

4. _____

5. Steve Zoltan held a 3-mo, 10% note for $750, dated March 31. He discounted the note at his bank on April 15 at a rate of $11\frac{1}{2}$%. The bank charges a collection fee of $2 or 0.1 percent, whichever is greater. What proceeds did Steve receive?

5. _____

Calculator Tip

When figuring discount on a calculator, change the time to a decimal.

Business Applications

6. On March 30, Green Thumb Lawn Services purchased a new power sprayer for $3,875. The vendor extended credit to Green Thumb by taking a 135-d note at $7\frac{1}{2}$%. On June 1, the vendor discounted the note at its bank at $9\frac{1}{4}$%. The bank charges a collection fee of $2.50 or 0.1 percent, whichever is greater. What were the proceeds of the note?

Answers

6. _____

7. On March 17, Daimler Automotive issued a 45-d sight draft in the amount of $215,000 with interest at $6\frac{3}{4}\%$ exact interest from the date of issue to the maturity date. The draft was accepted on June 11 and discounted the same day at $7\frac{1}{2}\%$ banker's interest. Compute the amount of proceeds the payee will receive.

8. Isaac Freeman is buying a new compressor priced at $1,652. Air Products will finance the purchase but compute the interest on an exact basis. Isaac's bank will provide the same financing with interest computed on the banker's basis. If Isaac signs a 72-d 11% note on November 11, will the bank or Air Products provide the lower financing cost? If Air Products discounted the note at $11\frac{3}{4}\%$ on December 1, what proceeds will it receive?

9. From which of these players would you be willing to take an interest-bearing note? Why?

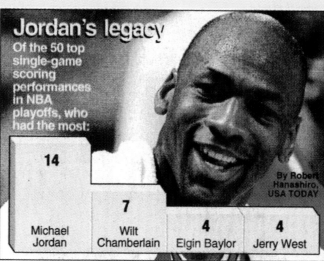

Jordan's legacy

Of the 50 top single-game scoring performances in NBA playoffs, who had the most:

14 — Michael Jordan
7 — Wilt Chamberlain
4 — Elgin Baylor
4 — Jerry West

Source: *Official NBA Guide 1993-94*

By Robert Hanashiro, USA TODAY

By Nick Galifianakis, USA TODAY

SKILLBUILDER **7.7**

Installment Purchases

Learning Outcomes

After completing this skillbuilder, you should be able to:

- Compute the cost of buying on time.
- Compute the real rate of interest on an installment purchase.

Computing the Cost of Buying on Time

Installment purchasing, or **buying on time,** is paying for an item by making a down payment and then making a series of payments, or installments, over a specified period of time. The usual practice is to charge interest for the privilege of making an installment purchase, although sometimes an installment purchase may cost no more than a cash purchase. The installment purchaser is generally charged various fees, such as an application fee for the loan, a credit investigation fee, and an insurance fee on the amount of the loan. These are added costs of buying on time.

To find the added cost of making an installment purchase, follow these steps.

1. Multiply the number of payments by the amount of each payment.
2. Add this product plus any fees or charges to the down payment. This total is the **total installment price**.
3. Subtract the cash purchase price from the total installment price.

The difference will be the added cost, if any, of buying on time.

MATH TIP

When figuring the total cost of financing, remember to include any fees related to buying on time.

Price Co. sells a color television for $450 cash. If the TV is purchased on an installment plan, a down payment of $45 and 15 monthly payments of $30.26 each are required. What is the added cost, if any, of buying the television on time?

SOLUTION

$453.90	Monthly payments: $15 \times \$30.26$
+ 45.00	Plus down payment
$498.90	Total installment price
− 450.00	Less cash price
$48.90	Added cost of buying on time

▶ Self-Check

The cash price of a copier is $1,300. If the copier is purchased on the installment plan, a down payment of $250 and 12 monthly payments of $104.60 are required. In addition, $10 is charged for a credit report. What is the added cost of buying on time?

Self-Check Answer

$215.20

Problems

1.

	Installment Terms			Installment Price	Cash Price	Cost of Buying on Time
Item	**Down Payment**	**Number of Payments**	**Monthly Payment**			
a. Word processor	None	12	$21.89	_____	$240.00	_____
b. CD player	$11.95	4	27.00	_____	119.95	_____
c. Stereo	None	15	20.77	_____	279.95	_____
d. Microwave	39.95	24	16.72	_____	379.95	_____
e. Cordless phone	19.99	9	18.17	_____	169.99	_____
f. Refrigerator	58.00	30	21.32	_____	579.95	_____

Answers

2. _____

3. _____

2. George Manley purchased a $319.95 dishwasher for his café on an installment purchase plan. The purchase plan required a down payment of $31.95 and weekly payments of $4.25 for 78 wk. What was the added cost of buying on time?

3. Lauren Black purchased a used car priced at $8,795 for Lauren's Catering Service. In order to purchase on the installment plan, she had to make a 20% down payment and pay the balance in 24 monthly payments of $358 each. How much more did it cost to buy the car on time?

Problems

4. Rohm's Furniture purchased a ZIP486D computer for $2,799. It paid 15% down and financed the balance over 24 mo at $123 per month. How much did it cost to finance the purchase?

4. _____

5. The Office Company purchased a 17,000-BTU air conditioner for a cash price of $397. It decided to finance the purchase with no money down and 12 monthly payments of only $37 per month. How much did it cost to finance the purchase?

5. _____

6. Ramon and Anita purchased five pieces of office furniture for their accounting business for $899, to be paid off with 18 payments of $57 per month. The payments include a $15 application fee and $10 for a credit report. How much did it cost to buy the furniture on time?

6. _____

7. Sam's Pizza bought a new car priced at $7,304. After paying $1,000 down, the monthly payments were $125 per month for 60 mo. What was the total financing charge for this car?

7. _____

8. Chamar's Carpets originally priced a 9- × 12-foot Persian carpet at $3,290. It was sale-priced at 40% off; Susan bought it by making 15 payments of $152 per month. How much more did Susan pay by buying the carpet on time?

8. _____

Problems

9. Tom Meyer's Appliances originally sold a 30-in. stereo TV at the regular price of $799. Sue-Ellen bargained the price down to $699.95. Her monthly payments were $52 per month for 15 mo. How much more did it cost her to finance the purchase?

10. Roger's Copies leased a copier for 36 mo. The lease agreement called for a payment of $125 per month for 36 mo after a $500 down payment. Each payment contained interest and other charges, which came to $23.02 per month. What was the original price of the copier?

Computing the Real Rate of Interest

Businesses that finance purchases need to be aware of the Truth-in-Lending Act. This act applies to credit transactions involving personal, family, or household loans not exceeding $25,000. The act requires that the lender state the **true annual rate of interest (APR)** and the **finance charge,** which is any additional amount paid as a result of buying on credit instead of paying cash. The rate may be stated as "$1\frac{1}{2}\%$ a month (18% annual rate)." This tells the borrower that a rate of 18 percent is being charged only on the amount of money actually owed.

   Add-on interest assumes that the borrower owes the entire sum of money over the period of the loan because the interest charged is computed for the entire period of the loan. Since a portion of the principal is being repaid with each payment, the borrower actually has the full use of the money for only a part of the period of the loan.

   Neither add-on interest nor additional charges give a true picture of the real rate of interest being paid. To find the real rate of interest for a purchase, use this formula.

$$\text{Real rate of interest} = \frac{2 \times 12^* \times \text{added cost of buying on time}}{\text{amount of loan} \times (\text{number of payments} + 1)}$$

* If payments are made monthly, this number is 12. If payments are made weekly, this number is 52.

Problems

A typewriter with a cash purchase price of $240 is sold on the installment purchase plan with no down payment. Add-on interest at an annual rate of 10% is charged to the purchaser, along with an application fee of $2.20, a credit check fee of $5, and insurance at a flat rate of $\frac{1}{2}\%$ of the amount of the loan.

If the typewriter is to be paid for in 1 y, what is the cost of buying on time, and what is the monthly payment? Find the real rate of interest.

SOLUTION

$240.00	Cash price and amount of loan
24.00	Plus interest: $240 × 0.10 × 1
7.20	Plus application fee and credit check
+ 1.20	Plus insurance charge: $240 × 0.005
$272.40	Total installment price
−240.00	Less cash price
$32.40	Added cost of buying on time
22.70	Monthly payment: $272.40 ÷ 12

$$\text{Real rate of interest} = \frac{2 \times 12^* \times \text{added cost of buying on time}}{\text{amount of loan} \times (\text{number of payments} + 1)}$$

$$= \frac{2 \times 12 \times \$32.40}{\$240 \times (12 + 1)} = \frac{24 \times \$32.40}{\$240 \times 13}$$

$$= \frac{\$777.60}{\$3,120} = 0.2492 \text{ or } 24.9\%$$

Self-Check

An office microwave was purchased with $39.95 down for 24 mo. The cash price was $379.95, and the monthly payment was $16.72. Find the real rate of interest.

Self-Check Answer

17.3%

Problems

11.

Item	Installment Terms				Real Rate of Interest
	Cash Price	Down Payment	Number of Payments	Monthly Payment	
a. Cordless phone	$169.99	$19.99	9	$18.17	_____
b. File cabinet	76.50	6.50	12	6.50	_____
c. Stereo	279.95	None	15	20.77	_____

Calculator Tip

You can calculate the real rate of interest in one series of steps by multiplying the three factors in the numerator, dividing by the amount of the loan, and then dividing by the number of payments plus 1.

12. In Problem 4, Rohm's Furniture purchased a computer for $2,799, paid 15% down, and financed the balance over 24 mo, with payments of $123 per month. What was the real rate of interest paid?

Answers

12. _____

13. Midland Publishing leased a copier. The lease cost $828.75 to finance $3,671.28 over 36 months. What real rate of interest did Midland pay on the lease?

13. _____

14. Tanya's Leather Goods sells a leather jacket for $285. It can be purchased for $51 per month over a 6-mo period. What real rate of interest is being paid?

14. _____

Problems

15. Eduardo's Lawn Stuff sells a lawn tractor for $900. It can be purchased for $19 per month over a 5-y period. What real rate of interest is charged?

16. Pets and More purchased a Teensie notebook computer with an 80-megabyte hard drive that sells for $1,399. The monthly payment for a 15-mo loan is $107.80 a month. (a) What is the cost of financing this purchase, and (b) what is the real rate of interest being paid?

Business Applications

17. Lonnie Gwynere borrowed $5,000 for 4 months in order to purchase inventory for her auto parts shop. The bank charged 9% interest for the period on a banker's basis. She is required to pay back principal plus interest in four equal monthly installments. (a) How much is each installment? (b) Is she really paying 9% interest?

18. Elizabeth purchased an electronic surveyor's transit for $12,875. She was required to pay 20% down, 5% add-on interest for 2 y, a loan application fee of $25, a credit-check fee of $10, and insurance on the purchase at a flat rate of $\frac{1}{2}$% of the loan. Compute the amount of Elizabeth's monthly payment to the nearest dollar.

Business Applications

19. Pali's Department store purchased a copy machine. (a) Compute the amount of the monthly payment of the copier if it sells at $4,000 with a 10% down payment. Interest at 11% is $313.50; insurance is $60; credit report and application fee is $35; and a mandatory service contract is $10 per month. The period of the loan is 18 mo. (b) What is the real rate of interest being paid?

Answers

19 a. _____

b. _____

20. If television viewing parallels movie-going, which 2 months of the year would have more TV installment sales?

20. _____

Movies hot in summer
Percentage of all movie-goers who attend in:

July 11.8%

12%
10%
8%
6%
4%

Jan.   Dec.

Source: *Frequent Moviegoer,*
The Interep Radio Store

By Cindy Hall and Elys A. McLean, USA TODAY

Student Notes

SKILLBUILDER 7.8

Computing Interest on Unpaid Balances

Learning Outcomes

After completing this skillbuilder, you should be able to:

- Compute the installment purchase price on a month-to month basis.
- Compute interest on the unpaid balance of an installment purchase.

Computing the Installment Purchase Price

Because each payment made on an installment purchase pays back a portion of the principal, the fairest basis is to base the interest on the unpaid balance. This can be done by computing the interest on the balance each month and then subtracting the payment to find the new balance.

The usual practice is for equal installment payments to be made over the period of the loan. To find the amount of each installment, the principal plus interest is divided by the number of payments.

MATH TIP

When monthly payments are not exact, remember to round up to the next whole cent.

EXAMPLE

A lawn mower was purchased on April 1 for $240 less a down payment of $60. The balance of $180 is to be paid off in equal monthly installments over a period of 3 mo with interest on the unpaid balance at 9%.

a. Prepare a table of the installment payments and balance due each month.

b. Find the amount of each payment if the loan is repaid in equal installments.

SOLUTION

a.

$$\text{Payment Toward Principal} = \text{Balance} \div \text{Number of Months}$$

$$\$60 = \$180 \div 3$$

$$\text{Interest Rate for Month} = \frac{\text{Interest Rate for Year}}{12}$$

$$= \frac{0.09}{12}$$

$$0.0075$$

$$\text{Interest for Month} = \text{Unpaid Balance} \times \text{Interest Rate for Month}$$

$$\$1.35 = \$180 \times 0.0075$$

Payment Date	Unpaid Balance	Interest for Month	Amount Due	Installment Payment	Balance Due
May 1	$180	$1.35	$181.35	$61.35	$120
June 1	120	.90	120.90	60.90	60
July 1	60	.45	60.45	60.45	0

b.

$180.00 Principal payments: $3 \times \$60$

   2.70 Plus interest owed: $1.35 + $0.90 + $0.45

$182.70 Total

$$\text{Monthly Payment Toward Loan} = \text{Principal and Interest Balance} \div \text{Number of Months}$$

$$\$60.90 = \$182.70 \div 3$$

▶ Self-Check

Mikovec Electric purchased a copier for $1500. Find the amount of each payment if it paid for the copier in three equal monthly installments and 12% interest was charged.

Self-Check Answer

$510

© by Glencoe/McGraw-Hill.

Problems

Find the amount due each month on the following installment sales if equal payments are made against the principal. Carry out the interest rate for each month to four decimal places.

1. Donaldson Video Court purchased a video game on the installment plan. The cash price was $159.99 less a down payment of $19.99. The balance was paid off over 4 mo with interest at 10% on the unpaid balance.

a. Complete this table to determine the balance due each month.

Payment Number	Unpaid Balance	Interest for Month	Amount Due	Installment Payment	Balance Due
1	_____	_____	_____	_____	_____
2	_____	_____	_____	_____	_____
3	_____	_____	_____	_____	_____
4	_____	_____	_____	_____	_____

b. What would the equal monthly payments for principal and interest be? (Round up your final answer to the next whole cent.)

Problems

2. Calorie Counting Cafeteria is replacing its central air conditioner at a cost of $799. After a down payment of $79, the balance will be paid in 6 monthly installments, with interest at 12% on the unpaid balance. What is the balance due each month?

Payment Number	Unpaid Balance	Interest for Month	Amount Due	Installment Payment	Balance Due
1					
2					
3					
4					
5					
6					

3. Roland replaced his heavy-duty mower at a price of $627. He was required to pay 25% down and will pay off the balance over 3 mo with interest at 11%. In addition, he was charged $15 for a credit report, which will be included in his monthly payments. What is the balance due each month?

Payment Number	Unpaid Balance	Interest for Month	Other Charges	Amount Due	Installment Payment	Balance Due
1						
2						
3						

Problems

4. The replacement cost for a checkout scanner is $1,323.53. The old unit has a trade-in value of $198.53. In addition, there is an insurance charge of $\frac{1}{2}$% of the amount of the loan. There is also a processing charge of $37.12. With interest charged at $9\frac{1}{2}$% for 9 mo, prepare a schedule to show how much is due each month and the balance.

Payment Number	Unpaid Balance	Interest for Month	Other Charges	Amount Due	Installment Payment	Balance Due
1	_____	_____	_____	_____	_____	_____
2	_____	_____	_____	_____	_____	_____
3	_____	_____	_____	_____	_____	_____
4	_____	_____	_____	_____	_____	_____
5	_____	_____	_____	_____	_____	_____
6	_____	_____	_____	_____	_____	_____
7	_____	_____	_____	_____	_____	_____
8	_____	_____	_____	_____	_____	_____
9	_____	_____	_____	_____	_____	_____

Problems

Using a Shortcut for Computing Interest on Unpaid Balances

Rather than computing the interest separately for each payment, you can use the following formula:

$$\text{Monthly payment toward loan} = \text{principal and interest balance} \div \text{number of months}$$

where the interest on the unpaid balances can be found by the formula:

$$\text{Interest on unpaid balances} = \frac{\text{first month's interest} \times (\text{number of payments} + 1)}{2}$$

When the monthly payment is not an exact number, the usual practice is to round up to the next whole cent. This will give a final payment that is slightly smaller than the other payments.

EXAMPLE

Use the formula to find the monthly payment on the lawn mower described in the first example.

$$\begin{array}{llll}
\textit{First Month's} & & & \\
\textit{Interest} & = \textit{Principal} & \times \textit{Rate} & \times \textit{Time} \\
\$1.35 & = \$180 & \times 0.09 & \times \frac{1}{12}
\end{array}$$

$$\text{Interest on unpaid balances} = \frac{\text{first month's interest} \times (\text{number of payments} + 1)}{2}$$

$$= \frac{\$1.35 \times (3 + 1)}{2}$$

$$= \frac{\$5.40}{2} = \$2.70$$

$$\begin{array}{llll}
\textit{Monthly Payment} & & \textit{Principal and} & \textit{Number of} \\
\textit{Toward Loan} & = & \textit{Interest Balance} & \div \quad \textit{Months}
\end{array}$$

$$\frac{\$182.70}{3} = \$60.90 = (180 + \$2.70) \div 3$$

 **Self-Check**

	Self-Check Answer
Find the monthly payment if $3,000 is to be repaid in 6 equal installments and the interest rate charged is 15% on the unpaid balance.	$521.88

Problems

Calculator Tip

Set the decimal-point selector at four and the round-off switch in the five/four position. Mentally round to the next highest cent.

5. Paul Graham bought a used delivery van for $4,250. He used his old compact car valued at $250, plus $750 cash, for the down payment. The balance is to be paid in 15 equal installments with interest at 8% on the unpaid balance. Compute (a) the interest on the unpaid balances and (b) the equal monthly payment for the purchase (round up your final answer to the next whole cent).

Answers

5. a.—————————

   b.—————————

Business Applications

6. Lorne's Livery purchased a new stretch limousine for $68,000. After trade-in, it obtained a loan for $34,000 at $8\frac{3}{4}$% for 7 y. Compute (a) the interest on the unpaid balances and (b) the equal monthly payment for the loan.

Answers

6. a.—————————

   b.—————————

7. Herb Neal was able to obtain a $5,200 loan at $12\frac{1}{2}$% for 15 mo, to install a new heating system in his restaurant. In addition, he is required to pay $86 insurance, inspection fees of $122, and a credit report charge of $15. What will be his monthly payments?

7. _____

8. Charlene Lopez bought a Mercedes diesel truck for her catering business. The truck has a price tag of $48,500, but with her late-model trade-in, she was able to reduce the price by 45%. The cost of customizing the interior is $12,850. She obtained a 3-y loan to cover the total amount at $9\frac{3}{4}$%. What was the amount of the monthly payment?

8. _____

9. You charge $65.00 on your credit card for dinner. When the bill arrives, you make a $10.00 payment. If your credit card charges 21% interest on the unpaid balance, how much do you owe next month?

9. _____

Where we eat takeout food

Home 53%
Work 14%
Car 19%
Other 14%

Source: Beef Industry Council

By Marcy E. Mullins, USA TODAY

SKILLBUILDER **7.9**

Computing Cash Discounts

After completing this skillbuilder, you should be able to:

- Compute cash discounts and net payments.
- Compute the balance due after a partial payment.
- Determine the cost of credit when a cash discount is involved.
- Compute single and multiple trade discounts.

Computing Cash Discounts and Net Payments

A **cash discount** is a deduction allowed on money owed in order to encourage prompt payment. The rate of discount and the time period allowed to take advantage of the cash discount are called the **terms.** Assume that the terms on an invoice are 3/10, 2/30, n/60. These payment terms provide two opportunities for the buyer to take advantage of a cash discount. The expression 3/10 means that a 3% discount can be taken if the bill is paid within 10 d. The second expression, 2/30, means that only a 2% discount can be taken if payment is made 11 to 30 d after the date of the invoice. More than one cash discount period may be indicated, but only one cash discount is allowed. The expression n/60 means that payment in full is due within 60 d of the invoice date. Cash discounts are never taken on service charges, such as freight, packing, insurance, and so on. If charges of this kind are included in the invoice, they must be deducted before the cash and trade discounts are taken, and then added back in, to find the total amount due.

Cash discount periods are sometimes counted from the end of the month in which the invoice is dated (EOM = end of month) or after the merchandise has been received (ROG = receipt of goods). With EOM, invoices dated from the first day of the month through the twenty-fifth day of the month have a cash discount period from the invoice date to 10 d after the end of that month. If the invoice is dated from the twenty-fifth to the end of the month, then an extra month is allowed for the cash discount period, with the due date being 10 d after the end of the following month. With both EOM and ROG, net payment is due 20 d after the last cash discount date, unless other terms are provided.

MATH TIP

Discount is sometimes given as a percent of the regular price. This is the discount rate. Discount rate × regular price = discount.

EXAMPLE

An invoice for $1,700, dated May 28, has terms of 3/10, 2/30, n/45. What is the last date on which each cash discount can be taken? When is the net payment due? What is the amount due in each instance?

SOLUTION

$$May\ 28 + 10\ d = June\ 7$$

$1,700	Amount of invoice
− 51	Less cash discount: $1,700 × 0.03
$1,649	Net payment

$$May\ 28 + 30\ d = June\ 27$$

$1,700	Amount of invoice
− 34	Less cash discount: $1,700 × 0.02
$1,666	Net payment

$$May\ 28 + 45\ d = July\ 12$$

$1,700	Amount of invoice
− 0	No cash discount allowed after June 27
$1,700	Net payment

EXAMPLE

Terms on an invoice dated August 10 are 3/15, n/45 ROG. The shipment was received on September 29. What is the last date on which the 3% cash discount can be taken, and what is the date for net payment?

SOLUTION

Since the goods were received on September 29, the discount can be taken for 15 d after that date, or until October 14. Net payment is due November 13.

EXAMPLE

An invoice is dated October 30 with terms of 3/10 EOM. What is the last date on which the 3% cash discount can be taken? What is the date for net payment?

SOLUTION

Since the invoice is dated after the twenty-fifth day of the month, we skip November and count 10 d into December for the cash discount period, which thus ends December 10. The net payment date is December 30.

Problems

1. Assume that these bills are paid within the cash discount period. Find the cash discount and the net payment for each invoice.

Amount of Invoice	Terms	Cash Discount	Net Payment
a. $ 875.00	5/20, n/30	_____	_____
b. $1,658.00	3/10, n/30	_____	_____

2. Find the cash discount, if any, and the net payment for these invoices.

Amount of Invoice	Terms	Date of Invoice	Date of Payment	Cash Discount	Net Payment
a. $3,800.00	3/10, n/30	October 13	October 16	_____	_____
b. 1,725.30	2/10, 1/30, n/60	November 20	December 7	_____	_____
c. 5,167.51	10/10, 3/30, n/60	June 15	July 16	_____	_____
d. 2,950.00	8/10, 4/30, n/60	March 23	April 24	_____	_____

Problems

3. Find the discount period, the cash discount, if any, and the net payment for these invoices.

Date of Invoice	Terms	Date Goods Received	Date of Payment	Discount Period	Amount of Invoice	Cash Discount	Net Payment
a. Nov. 21	2/10, n/30 EOM	Nov. 23	Dec. 8	_____	$ 727.80	_____	_____
b. Sept. 16	3/15, n/30 EOM	Sept. 22	Oct. 14	_____	1,464.00	_____	_____
c. June 6	3/10, n/30 ROG	June 12	June 20	_____	2,748.00	_____	_____
d. Dec. 11	4/10, n/30 ROG	Dec. 18	Dec. 20	_____	482.00	_____	_____
e. Jan. 6	2/10, n/30 EOM	Jan. 20	Feb. 5	_____	4,226.10	_____	_____
f. Aug. 19	1/10, n/30 EOM	Aug. 30	Sept. 11	_____	864.80	_____	_____

Answers

4. Jameson Products received a shipment of electronic parts on October 25. The invoice, in the amount of $985.50, was dated October 20. The cash discount terms were 3/10, n/45 EOM.

a. What is the last day on which a cash discount may be taken?

b. What amount must be paid then?

4. a. _____

b. _____

5. A shipment invoiced at $3,650 was received by Gable Inc. on May 17. The invoice was dated April 29, with terms of 3/10, n/30 ROG. If payment is made on May 26, how much would Gable Inc. pay?

5. _____

6. In the previous problem, if the invoiced amount of $3,650 included shipping charges of $112.50, insurance charges of $36.50, and special packing charges of $25, how much would Gable Inc. pay on May 26?

6. _____

Problems

7. Louise Kurtz received a shipment from Boone Electronics in the amount of $12,500. The invoice was dated February 29 with terms of 5/15 EOM. (a) What is the last day on which the cash discount may be taken? (b) How much will she pay? (c) What is the date for net payment?

Answers

7. a. ──────────

   b. ──────────

   c. ──────────

Computing Balance Due After a Partial Payment

When approved by the seller, the buyer can make payment of only a portion of the invoice and still receive a cash discount on the amount paid. Any remaining balance paid after the cash discount period is paid net. To determine the amount of credit to be allowed for the partial payment, subtract the rate of the cash discount from 100% and divide the amount of the partial payment by the difference.

EXAMPLE

An invoice in the amount of $12,000, dated May 17, has terms of 3/10, n/30. If a partial payment of $5,000 is made on May 27, what amount would be payable on the net date?

SOLUTION

$12,000.00   Amount of invoice
            Cash discount rate = 3%; 100% − 3% = 97%
− 5,154.64   Partial payment: $5,000 ÷ 0.97
$ 6,845.36   Balance due on net date

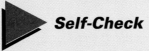 **Self-Check**

Find the balance due on the net date for an invoice in the amount of $20,000 with the terms of 5/15, n/45, when a partial payment in the amount of $10,000 is made within the 15-d cash discount period.

Self-Check Answer

$9,473.68

8. An invoice in the amount of $7,200 is dated August 12. Terms are 5/10, 3/30, n/60. A payment in the amount of $3,600 is made on August 23. (a) What is the net payment date, and (b) how much is due at that time?

8. a._____

b._____

9. Brenda Freidman purchased a new wood-turning lathe for $1,857. The invoice was dated January 12, with terms of 2/20, n/30. Brenda paid $1,000 on February 2. What amount was due on the net date?

9. _____

10. Henrietta Chen received a shipment in the amount of $28,300. Included in the amount were $782 for freight, $152 for packing, and $78 for insurance. The invoice was dated May 30, with terms of 3/15, 1/30, n/45. Henrietta made a partial payment of $10,000 on June 12 and a further payment of $8,000 on June 29. (a) What is the net date, and (b) how much is due at that time?

10. a._____

b._____

11. James Purdy purchased new lawn furniture off-season for $2,251. The invoice was dated October 27 with terms of 5/15, n/25. James paid $1,100 on November 10. How much did he pay on the net date?

11. _____

Problems

Determining the Cost of Credit

Many businesses have cash-flow problems; although they are otherwise financially solvent, they may not have immediate access to the cash needed to pay current bills, particularly to take advantage of cash discount provisions. A common practice is to obtain a short-term loan to cover the period from the last cash discount date to the net payment date. This practice works because the borrower of a loan is charged interest, on an annual basis, for the period the money is actually used, whereas the cash discount is a flat percent for that specific time period.

EXAMPLE

An invoice in the amount of $5,000 is dated May 1, with terms of 3/10, n/30. In order to take advantage of the cash discount, it is necessary to obtain a 9% loan for the period from the last cash discount date to the net payment date. Will it be profitable to take out such a loan?

SOLUTION

$5,000	Amount of invoice
− 150	Cash discount: $5,000 × 0.03
$4,850	Amount of loan needed
	Cost of loan: $4,850 × 0.09 × $\frac{20}{360}$ = $24.25
$150.00	Amount of cash discount
− 24.25	Cost of loan
$125.75	Amount saved

 **Self-Check**

Myrtle Ellis received a shipment of VCR cassettes invoiced at $8,700. The invoice is dated March 27, with terms of 2/15, n/30. Myrtle has an account receivable in the amount of $15,000 due April 15. In order to take advantage of the cash discount, she has been quoted a 12% loan rate from her bank. Should Myrtle borrow the money necessary to pay within the cash discount period? How much will it cost her?

Self-Check Answer

Yes, $131.37 net savings

12. Fred Hergesheimer received a $4,200 shipment on August 10. The invoice was dated August 3, with terms of 5/10, 2/20, n/45. On August 14 he obtained a loan at $10\frac{1}{2}\%$ to pay for the shipment. (a) How much should Fred borrow? (b) Will he save any money doing so?

13. Darrell Pfeil purchased 850 cases of canned fruit for $16,285. The invoice was dated November 28, with terms of 4/10 EOM. On December 8, he obtained a loan at 11% to pay for the shipment. (a) How much did he borrow, and (b) what was the cost of the loan? (c) What were his savings, if any?

Calculator Tip

To determine the net payment, subtract or use the memory key. Use the percent key to find the discount. You can also use the percent key when dividing by a percent.

Computing Trade Discounts

The price at which an article is sold is the **list price** or **catalog price.** A **trade discount** is a deduction from the list price of an article. A trade discount is offered to qualified purchasers, such as retailers or others who purchase in large quantities. The cost to the retailer after the trade discount is taken is the **net price.**

The discount is expressed as a percent, for example, 15% off the list price. A series of discounts may be offered. When computing the amount of a series of trade discounts, the order in which the discounts are applied will not affect the results.

The net price can also be computed directly. First deduct the trade discount percent from 100% then multiply the list or discounted prices by the remainder.

Problems

MATH TIP

When working discount problems, you can use either the discount rate or its complement (100% − the discount rate).

EXAMPLE

What is the net price of a stereo with a list price of $380, less trade discounts of 10% and 5%?

SOLUTION

$$\begin{array}{ccccc}
\textit{First Trade} & = & \textit{List} & \times & \textit{First Trade} \\
\textit{Discount} & & \textit{Price} & & \textit{Discount Rate} \\
\$38 & = & \$380 & \times & 0.10
\end{array}$$

$$\begin{array}{rl}
\$380 & \text{List price} \\
-\quad 38 & \text{First trade discount} \\
\hline
\$342 & \text{Discounted price}
\end{array}$$

$$\begin{array}{ccccc}
\textit{Second Trade} & = & \textit{Discounted} & \times & \textit{Second Trade} \\
\textit{Discount} & & \textit{Price} & & \textit{Discount Rate} \\
\$17.10 & = & \$342 & \times & 0.05
\end{array}$$

$$\begin{array}{rl}
\$342.00 & \text{Discounted price} \\
-\quad 17.10 & \text{Second trade discount} \\
\hline
\$324.90 & \text{Net price}
\end{array}$$

To solve using the direct method, subtract each percent from 100%:

$$100\% - 10\% = 90\% \qquad 100\% - 5\% = 95\%$$
$$\$380 \times 0.90 = \$342$$
$$\$342 \times 0.95 = \$324.90$$

 **Self-Check**

Self-Check Answer

What is the net price of a compressor with a list price of $590, less trade discounts of 15% and 5%?

$476.43

Problems

14. Find the trade discount and net price for the following.

	Trade Discount	Net Price
a. $50 less 20%	_____	_____
b. $180 less 25% and 15%	_____	_____
c. $900 less 48% and 22%	_____	_____

15. Use the direct method to compute each net price.
a. $750 less 10%
b. $2,250 less 14% and 20%
c. $400 less 5%, 15%, and 25%

15. **a.** _____
    b. _____
    c. _____

16. Use the direct method to compute the list price from the net price in each case.

Net Price	Trade Discount Rate
a. $720	10%
b. $646	5% and 15%
c. $28.56	20%, 30%, and 40%

16. **a.** _____
    b. _____
    c. _____

Problems

17. Mary Roman received a shipment of copper tubing on June 2. The invoice was dated May 31, with trade discounts of 15%, 5%, and 10% and terms of 5/10, n/30. Because of a printer error the invoice did not show the list price for the tubing, only the total and a listing of miscellaneous charges. The total for the shipment was $9,376.82. There were charges of $218 for freight, $57 for packing, and $90.12 for insurance. Mary wanted to take advantage of the cash discount period, which had only 8 d to run but wanted to be sure that she received the appropriate trade discounts. What was (a) the list price for the tubing purchased, and what was (b) the final amount which she paid?

Answers

17. a. _____

b. _____

18. A cash discount on travel expenses was offered if a companion traveled with the business person. If the original bill is $85 per night for a room, with a 30% discount on the second room, how much do 2 rooms cost each night?

18. _____

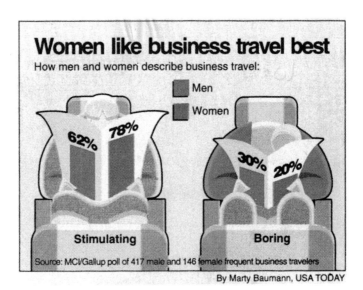

Women like business travel best

How men and women describe business travel:

■ Men
■ Women

62% 78%

30% 20%

Stimulating

Boring

Source: MCI/Gallup poll of 417 male and 146 female frequent business travelers

By Marty Baumann, USA TODAY

Business Applications

19. Prendevil Products received a shipment invoiced at $5,500, including $300 freight. The invoice is dated April 15, with terms of 4/10, 3/30, n/60. Prendevil paid for the shipment on April 29. How much did they pay?

19. _____

20. Peter's Burgers purchased a deluxe gas grill for $563, including shipping and insurance charges of $31.85. The invoice is dated February 21, 1996, with terms of 3/15, n/45. What is the (a) last date to obtain the cash discount and the (b) date for net payment? (c) If Peter takes the cash discount, how much will he pay for the grill?

20. a. _____

b. _____

c. _____

21. Melodie Hunt received a $7,800 shipment on October 12. The invoice is dated September 29, with terms of 5/15, 3/30, 1/45, n/60. If payment is made on November 14, how much will she pay?

21. _____

Business Applications

22. If a $5,000 travel package to the 1996 Olympics was discounted 20%, what would be the cost?

Answers

22. ——————

Olympian ticket totals

Tickets for the 1996 Atlanta Summer Olympics go on sale in the spring of 1995. Tickets available and sold:

	Tickets (in millions)	Percent sold
1984 Los Angeles	6.9	83%
1988 Seoul	4.3	74%
1992 Barcelona	3.9	68%
1996 Atlanta¹	10.2	61.5%

1 – projected

Source: Atlanta Committee for the Olympic Games

Days to opening ceremony: 795
For more information: 1-404-224-1996

By Marcy E. Mullins, USA TODAY

Student Notes

——
——
——
——
——
——
——
——
——
——
——
——
——
——
——
——
——

SKILLBUILDER 7.10

Changing a Discount Series to a Single-Discount Equivalent

Learning Outcomes

After completing this skillbuilder, you should be able to:

■ Convert a series of discounts to a single-discount equivalent.
■ Determine the net price directly from the list price.

Finding the Single-Discount Equivalent

A series of discounts can be converted to a single discount that is equivalent to the discount series. A **single-discount equivalent** is particularly useful when the same series of discounts is used frequently. The single-discount equivalent is *not* the sum of the original discounts because each discount after the first one is based on the **remainder** after the previous discount has been deducted.

MATH TIP

Remember to use the *complement* of the given discount rate when finding the single-discount equivalent. The complement of a percent is 100% minus that percent.

EXAMPLE

Determine the single-discount equivalent of the series 20%, 10%, and 5%.

SOLUTION

To find the single-discount equivalent, subtract each discount rate from 100%. Then find the product of these differences.

$$\begin{array}{ccccc} 100\% & & 100\% & & 100\% \\ -\ 20\% & & -\ 10\% & & -\ 5\% \\ \hline 80\% & \times & 90\% & \times & 95\% = 68.4\% \end{array}$$

Finally, subtract the result from 100% to get the single-discount equivalent.

$$100\% - 68.4\% = 31.6\% \text{ single-discount equivalent}$$

 **Self-Check**

Find the single-discount equivalent for a series discount of 10%, 20%, 15%.

Self-Check Answer

__38.8%__

Problems

Find the single-discount equivalent for each series discount. Round decimals to four decimal places, where necessary.

1. 30%, 20%, 10%

2. 25%, 10%

3. 15%, 25%, 5%, 10%

4. $33\frac{1}{3}$%, 15%, 5%

5. 50%, $12\frac{1}{2}$%, $6\frac{1}{4}$%

6. 40%, 20%, 10%

7. 60%, 30%, 15%

8. 15%, $7\frac{1}{2}$%, 5%

9. 40%, 15%, $2\frac{1}{2}$%, 10%

10. 35%, 15%, 5%, 40%, 10%

1. _____

2. _____

3. _____

4. _____

5. _____

6. _____

7. _____

8. _____

9. _____

10. _____

Finding Net Price

The **net price** is the amount to be charged for an item after all discounts have been applied. To find the net price, the discounts can be subtracted from the list price or a table such as the one shown on page 394 can be used. The horizontal and vertical rate columns list different discounts. Because the order in which discounts are taken does not affect the total discount, there may be several different ways to find the net of $1 (the net price of an article that originally cost $1).

For example, the discount series 10%, 5%, and $2\frac{1}{2}$% can be found in two ways. Read down the left hand rate column until two of the rates in the series are found. If the 5–$2\frac{1}{2}$% row is used, read across this row to the third rate in the series, 10%. The net of $1 for this series is 0.8336. Or, read down further on the left hand column to the 10–$2\frac{1}{2}$% row. Read across this row to the 5% column. The net of $1 is also 0.8336. After the net of $1 is found, multiply it by the total price of the goods purchased to find the net price.

Problems

An item with a list price of $320 is subject to discounts of 10%, 20%, and $2\frac{1}{2}$%. Find the single-discount equivalent for the item and the net price.

SOLUTION

Read down the rate column until two of the discount rates in the series are found: $10-2\frac{1}{2}$%. Read across this row to the 20% column. The factor at the intersection of this column and row is 0.702, which means that for each dollar subject to discounts of 10%, 20%, and $2\frac{1}{2}$%, only $0.702 would be paid.

Multiply this factor (0.702) by the total price ($320). The net price is $224.64. To find the single-discount equivalent, change 0.702 to a percent and subtract it from 100%.

$$100\% - 70.2\% = 29.8\% \text{ single-discount equivalent}$$

or

$$\begin{array}{ccc} 100\% & 100\% & 100\% \\ -\ \underline{\ 10\%} & -\ \underline{\ 20\%} & -\ \underline{2\frac{1}{2}\%} \\ 90\% \ \times & 80\% \times & 97\frac{1}{2}\% = 70.2\% \text{ Net factor} \end{array}$$

$$100\% - 70.2\% = 29.8\% \text{ Single-discount equivalent}$$

$$\$320 \times 70.2\% = \$224.64 \text{ Net price}$$

or

$$\begin{array}{l} \$320.00 \times 10\% = \$32.00 \\ -\ \underline{\ 32.00} \\ \$288.00 \times 20\% = \$57.60 \\ -\ \underline{\ 57.60} \\ \$230.40 \times 2\frac{1}{2}\% = \$5.76 \\ -\ \underline{\ 5.76} \\ \$224.64 \text{ Net price} \end{array}$$

Any one of these methods can be used to find the net price when a series of discounts is given.

 **Self-Check**

An item with a list price of $800 is subject to discounts of 20%, 10%, and $7\frac{1}{2}$%.

a. Find the single-discount equivalent for this item.
b. Find the net price of the item.

Self-Check Answers

a. _____33.4%_____

b. _____$532.80_____

Problems

Rate	5%	$7\frac{1}{2}$%	10%	15%	20%	25%	30%	$33\frac{1}{3}$%	40%	50%
2%	0.931	0.9065	0.882	0.833	0.784	0.735	0.686	0.6534	0.588	0.49
$2\frac{1}{2}$%	0.9263	0.9019	0.8775	0.8288	0.78	0.7313	0.6825	0.65	0.585	0.4875
5%	0.9025	0.8788	0.855	0.8075	0.76	0.7125	0.665	0.6334	0.57	0.475
5–$2\frac{1}{2}$%	0.8799	0.8568	0.8336	0.7873	0.741	0.6947	0.6484	0.6175	0.5558	0.4631
$7\frac{1}{2}$%	0.8788	0.8556	0.8325	0.7863	0.74	0.6938	0.6475	0.6167	0.555	0.4625
$7\frac{1}{2}$–5%	0.8348	0.8128	0.7909	0.7469	0.703	0.6591	0.6151	0.5859	0.5273	0.4394
10%	0.855	0.8325	0.81	0.765	0.72	0.675	0.63	0.60	0.54	0.45
10–$2\frac{1}{2}$%	0.8336	0.8117	0.7898	0.7459	0.702	0.6581	0.6143	0.585	0.5265	0.4388
10–5%	0.8123	0.7909	0.7695	0.7268	0.684	0.6413	0.5985	0.57	0.513	0.4275
10–5–$2\frac{1}{2}$%	0.7919	0.7711	0.7503	0.7086	0.6669	0.6252	0.5835	0.5558	0.5002	0.4168
10–10%	0.7695	0.7493	0.729	0.6885	0.648	0.6075	0.567	0.54	0.486	0.405
10–10–5%	0.7310	0.7118	0.6926	0.6541	0.6156	0.5771	0.5387	0.513	0.4617	0.3848
20–5%	0.722	0.703	0.684	0.646	0.608	0.57	0.532	0.5067	0.456	0.38
20–10%	0.684	0.666	0.648	0.612	0.576	0.54	0.504	0.48	0.432	0.36
25%	0.7125	0.6938	0.675	0.6375	0.60	0.5625	0.5250	0.50	0.45	0.375
25–5%	0.6769	0.6591	0.6413	0.6056	0.57	0.5344	0.4988	0.475	0.4275	0.3563
25–10%	0.6413	0.6244	0.6075	0.5738	0.54	0.5063	0.4725	0.45	0.405	0.3375
25–10–5%	0.6092	0.5932	0.5771	0.5451	0.513	0.4809	0.4489	0.4275	0.3848	0.3206

Calculator Tip

Use multifactor multiplication to determine the single-discount equivalent.

Use the table above to find the single-discount equivalent for each discount series.

	Discount Series	*Single-Discount Equivalent*
11.	25%, $7\frac{1}{2}$%	_____
12.	15%, 10%, 5%, $2\frac{1}{2}$%	_____
13.	10%, 40%, 10%, 5%	_____
14.	50%, 25%	_____
15.	10%, 5%, 10%	_____
16.	$7\frac{1}{2}$%, 5%, $7\frac{1}{2}$%	_____
17.	$33\frac{1}{3}$%, 10%, 5%, $2\frac{1}{2}$%	_____
18.	25%, 40%, 5%	_____
19.	40%, 10%, 5%, $2\frac{1}{2}$%	_____
20.	$33\frac{1}{3}$%, 10%, 5%, 25%	_____

Business Applications

21. Bertram Brothers offered their product subject to discounts of $7\frac{1}{2}$%, 5%, and 15%. Peach Brothers offer a competing product with discounts of 10%, 5%, $2\frac{1}{2}$%, and 10%. Both sets of discounts total $27\frac{1}{2}$%. Who offers the better price?

22. Highlights Unlimited received the following order:

    10 doz pens @ $13.58 list; discounts of 30%, 20%, 10%
    25 doz writing pads @ $15.87 list; discounts of 50%, 25%, 10%, 5%
    4 boxes computer paper @ $49.85 list; discounts of 50%, 20%, 10%
    12 packages facsimile paper @ $49.50 list; discounts of 50%, 10%, 10%, 5%

a. What is the single-discount equivalent for each item?
b. What is the total net price for this order?

23. Ruth Kerlin was billed $299.25 for a fax machine with a list price of $700. The unit was advertised with discounts of 50%, 20%, and 10%. (a) Was Ruth billed correctly? (b) If not, what should the net price have been? (c) What is the single-discount equivalent of the original billing?

Business Applications

24. An item originally priced at $5.99 is discounted 20% for a grand opening sale. The manufacturer offers a discount coupon for 10% off the discount price. What is the single-discount equivalent, and how much did the $5.99 item cost?

Answers

24. _____

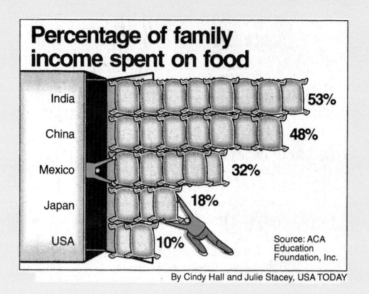

Percentage of family income spent on food

India 53%

China 48%

Mexico 32%

Japan 18%

USA 10%

Source: ACA Education Foundation, Inc.

By Cindy Hall and Julie Stacey, USA TODAY

Student Notes

APPENDIX

If the wages are–		And the number of withholding allowances claimed is—										
At least	But less than	0	1	2	3	4	5	6	7	8	9	10
		The amount of income tax to be withheld is—										
$0	$50	$0	$0	$0	$0	$0	$0	$0	$0	$0	$0	$0
50	55	1	0	0	0	0	0	0	0	0	0	0
340	350	44	38	31	24	17	11	4	0	0	0	0
350	360	46	39	32	26	19	12	5	0	0	0	0
360	370	47	41	34	27	20	14	7	0	0	0	0
370	380	49	42	35	29	22	15	8	2	0	0	0
380	390	50	44	37	30	23	17	10	3	0	0	0
390	400	52	45	38	32	25	18	11	5	0	0	0
400	410	53	47	40	33	26	20	13	6	0	0	0
410	420	55	48	41	35	28	21	14	8	1	0	0
420	430	56	50	43	36	29	23	16	9	2	0	0
430	440	58	51	44	38	31	24	17	11	4	0	0
440	450	59	53	46	39	32	26	19	12	5	0	0
450	460	61	54	47	41	34	27	20	14	7	0	0
460	470	64	56	49	42	35	29	22	15	8	1	0
470	480	67	57	50	44	37	30	23	17	10	3	0
480	490	70	59	52	45	38	32	25	18	11	4	0
490	500	73	60	53	47	40	33	26	20	13	6	0
500	510	75	63	55	48	41	35	28	21	14	7	1
510	520	78	66	56	50	43	36	29	23	16	9	2
520	530	81	68	58	51	44	38	31	24	17	10	4
530	540	84	71	59	53	46	39	32	26	19	12	5
540	550	87	74	61	54	47	41	34	27	20	13	7
550	560	89	77	64	56	49	42	35	29	22	15	8
560	570	92	80	67	57	50	44	37	30	23	16	10
570	580	95	82	70	59	52	45	38	32	25	18	11
580	590	98	85	73	60	53	47	40	33	26	19	13

If the wages are–		And the number of withholding allowances claimed is—										
At least	But less than	0	1	2	3	4	5	6	7	8	9	10
		The amount of income tax to be withheld is—										
$0	$125	$0	$0	$0	$0	$0	$0	$0	$0	$0	$0	$0
125	130	1	0	0	0	0	0	0	0	0	0	0
440	450	49	42	35	29	22	15	8	1	0	0	0
450	460	50	44	37	30	23	16	10	3	0	0	0
460	470	52	45	38	32	25	18	11	4	0	0	0
470	480	53	47	40	33	26	19	13	6	0	0	0
480	490	55	48	41	35	28	21	14	7	1	0	0
490	500	56	50	43	36	29	22	16	9	2	0	0
500	510	58	51	44	38	31	24	17	10	4	0	0
510	520	59	53	46	39	32	25	19	12	5	0	0
520	530	61	54	47	41	34	27	20	13	7	0	0
530	540	62	56	49	42	35	28	22	15	8	1	0
540	550	64	57	50	44	37	30	23	16	10	3	0
550	560	65	59	52	45	38	31	25	18	11	4	0
560	570	67	60	53	47	40	33	26	19	13	6	0
570	580	68	62	55	48	41	34	28	21	14	7	1
580	590	70	63	56	50	43	36	29	22	16	9	2
590	600	71	65	58	51	44	37	31	24	17	10	4
600	610	73	66	59	53	46	39	32	25	19	12	5
610	620	74	68	61	54	47	40	34	27	20	13	7
620	630	76	69	62	56	49	42	35	28	22	15	8
630	640	77	71	64	57	50	43	37	30	23	16	10
640	650	79	72	65	59	52	45	38	31	25	18	11
650	660	80	74	67	60	53	46	40	33	26	19	13
660	670	82	75	68	62	55	48	41	34	28	21	14
670	680	83	77	70	63	56	49	43	36	29	22	16
680	690	85	78	71	65	58	51	44	37	31	24	17
690	700	86	80	73	66	59	52	46	39	32	25	19
700	710	88	81	74	68	61	54	47	40	34	27	20
710	720	89	83	76	69	62	55	49	42	35	28	22
720	730	91	84	77	71	64	57	50	43	37	30	23
730	740	92	86	79	72	65	58	52	45	38	31	25

Time	$2\frac{1}{2}\%$	3%	$3\frac{1}{2}\%$	4%	$4\frac{1}{2}\%$	5%	$5\frac{1}{2}\%$	6%	$6\frac{1}{2}\%$	7%
1 day	.0069	.0083	.0097	.0111	.0125	.0139	.0153	.0167	.0181	.0194
2 days	.0139	.0167	.0194	.0222	.0250	.0278	.0306	.0333	.0361	.0389
3 days	.0208	.0250	.0292	.0333	.0375	.0417	.0458	.0500	.0542	.0583
4 days	.0278	.0333	.0389	.0444	.0500	.0556	.0611	.0667	.0722	.0778
5 days	.0347	.0417	.0486	.0556	.0625	.0694	.0764	.0833	.0903	.0972
6 days	.0417	.0500	.0583	.0667	.0750	.0833	.0917	.1000	.1083	.1167
7 days	.0486	.0583	.0681	.0778	.0875	.0972	.1069	.1167	.1264	.1361
8 days	.0556	.0667	.0778	.0889	.1000	.1111	.1222	.1333	.1444	.1556
9 days	.0625	.0750	.0875	.1000	.1125	.1250	.1375	.1500	.1625	.1750
10 days	.0694	.0833	.0972	.1111	.1250	.1389	.1528	.1667	.1806	.1944
11 days	.0764	.0917	.1069	.1222	.1375	.1528	.1681	.1833	.1986	.2139
12 days	.0833	.1000	.1167	.1333	.1500	.1667	.1833	.2000	.2167	.2333
13 days	.0903	.1083	.1264	.1444	.1625	.1806	.1986	.2167	.2347	.2528
14 days	.0972	.1167	.1361	.1556	.1750	.1944	.2139	.2333	.2528	.2722
15 days	.1042	.1250	.1458	.1667	.1875	.2083	.2292	.2500	.2708	.2917
16 days	.1111	.1333	.1556	.1778	.2000	.2222	.2444	.2667	.2889	.3111
17 days	.1181	.1417	.1653	.1889	.2125	.2361	.2597	.2833	.3069	.3306
18 days	.1250	.1500	.1750	.2000	.2250	.2500	.2750	.3000	.3250	.3500
19 days	.1319	.1583	.1847	.2111	.2375	.2639	.2903	.3167	.3431	.3694
20 days	.1389	.1667	.1944	.2222	.2500	.2778	.3056	.3333	.3611	.3889
21 days	.1458	.1750	.2042	.2333	.2625	.2917	.3208	.3500	.3792	.4083
22 days	.1528	.1833	.2139	.2444	.2750	.3056	.3361	.3667	.3972	.4278
23 days	.1597	.1917	.2236	.2556	.2875	.3194	.3514	.3833	.4153	.4472
24 days	.1667	.2000	.2333	.2667	.3000	.3333	.3667	.4000	.4333	.4667
25 days	.1736	.2083	.2431	.2778	.3125	.3472	.3819	.4167	.4514	.4861
26 days	.1806	.2167	.2528	.2889	.3250	.3611	.3972	.4333	.4694	.5056
27 days	.1875	.2250	.2625	.3000	.3375	.3750	.4125	.4500	.4875	.5250
28 days	.1944	.2333	.2722	.3111	.3500	.3889	.4278	.4667	.5056	.5444
29 days	.2014	.2417	.2819	.3222	.3625	.4028	.4431	.4833	.5236	.5639
1 month	.2083	.2500	.2917	.3333	.3750	.4167	.4583	.5000	.5417	.5833
2 months	.4167	.5000	.5833	.6667	.7500	.8333	.9167	1.0000	1.0833	1.1667
3 months	.6230	.7500	.8750	1.0000	1.1250	1.2500	1.3750	1.5000	1.6250	1.7500
4 months	.8333	1.0000	1.1667	1.3333	1.5000	1.6667	1.8333	2.0000	2.1667	2.3333
5 months	1.0417	1.2500	1.4583	1.6667	1.8750	2.0833	2.2917	2.5000	2.7083	2.9160
6 months	1.2500	1.5000	1.7500	2.0000	2.2500	2.5000	2.7500	3.0000	3.2500	3.5070

Day of Month	January	February	March	April	May	June	July	August	September	October	November	December	Day of Month
1	1	32	60	91	121	152	182	213	244	274	305	335	1
2	2	33	61	92	122	153	183	214	245	275	306	336	2
3	3	34	62	93	123	154	184	215	246	276	307	337	3
4	4	35	63	94	124	155	185	216	247	277	308	338	4
5	5	36	64	95	125	156	186	217	248	278	309	339	5
6	6	37	65	96	126	157	187	218	249	279	310	340	6
7	7	38	66	97	127	158	188	219	250	280	311	341	7
8	8	39	67	98	128	159	189	220	251	281	312	342	8
9	9	40	68	99	129	160	190	221	252	282	313	343	9
10	10	41	69	100	130	161	191	222	253	283	314	344	10
11	11	42	70	101	131	162	192	223	254	284	315	345	11
12	12	43	71	102	132	163	193	224	255	285	316	346	12
13	13	44	72	103	133	164	194	225	256	286	317	347	13
14	14	45	73	104	134	165	195	226	257	287	318	348	14
15	15	46	74	105	135	166	196	227	258	288	319	349	15
16	16	47	75	106	136	167	197	228	259	289	320	350	16
17	17	48	76	107	137	168	198	229	260	290	321	351	17
18	18	49	77	108	138	169	199	230	261	291	322	352	18
19	19	50	78	109	139	170	200	231	262	292	323	353	19
20	20	51	79	110	140	171	201	232	263	293	324	354	20
21	21	52	80	111	141	172	202	233	264	294	325	355	21
22	22	53	81	112	142	173	203	234	265	295	326	356	22
23	23	54	82	113	143	174	204	235	266	296	327	357	23
24	24	55	83	114	144	175	205	236	267	297	328	358	24
25	25	56	84	115	145	176	206	237	268	298	329	359	25
26	26	57	85	116	146	177	207	238	269	299	330	360	26
27	27	58	86	117	147	178	208	239	270	300	331	361	27
28	28	59	87	118	148	179	209	240	271	301	332	362	28
29	29	—	88	119	149	180	210	241	272	302	333	363	29
30	30	—	89	120	150	181	211	242	273	303	334	364	30
31	31	—	90	—	151	—	212	243	—	304	—	365	31

NOTE: For leap year, one day must be added to number of days after February 28.

Period	$\frac{1}{3}\%$	$\frac{1}{2}\%$	$\frac{2}{3}\%$	$\frac{3}{4}\%$	1%	$1\frac{1}{4}\%$	$1\frac{1}{2}\%$
1	1.003333	1.005000	1.006666	1.007500	1.010000	1.012500	1.015000
2	1.006677	1.010025	1.013377	1.015056	1.020100	1.025156	1.030225
3	1.010033	1.015075	1.020133	1.022669	1.030301	1.037970	1.045678
4	1.013400	1.020150	1.026934	1.030339	1.040604	1.050945	1.061363
5	1.016778	1.025251	1.033780	1.038066	1.051010	1.064082	1.077284
6	1.020167	1.030377	1.040672	1.045852	1.061520	1.077383	1.093443
7	1.023567	1.035529	1.047610	1.053696	1.072135	1.090850	1.109844
8	1.026979	1.040707	1.054594	1.061598	1.082856	1.104486	1.126492
9	1.030403	1.045910	1.061625	1.069560	1.093685	1.118292	1.143389
10	1.033837	1.051140	1.068702	1.077582	1.104622	1.132270	1.160540
11	1.037283	1.056395	1.075827	1.085664	1.115668	1.146424	1.177948
12	1.040741	1.061677	1.082999	1.093806	1.126825	1.160754	1.195618
13	1.044210	1.066986	1.090219	1.102010	1.138093	1.175263	1.213552
14	1.047691	1.072321	1.097487	1.110275	1.149474	1.189954	1.231755
15	1.051183	1.077682	1.104804	1.118602	1.160968	1.204829	1.250232
16	1.054687	1.083071	1.112169	1.126992	1.172578	1.219889	1.268985
17	1.058203	1.088486	1.119584	1.135444	1.184304	1.235138	1.288020
18	1.061730	1.093928	1.127047	1.143960	1.196147	1.250577	1.307340
19	1.065269	1.099398	1.134561	1.152540	1.208108	1.266209	1.326950
20	1.068820	1.104895	1.142125	1.161184	1.220190	1.282037	1.346854
21	1.072383	1.110420	1.149739	1.169893	1.232391	1.298062	1.367057
22	1.075957	1.115972	1.157404	1.178667	1.244715	1.314288	1.387563
23	1.079544	1.121552	1.165120	1.187507	1.257163	1.330717	1.408377
24	1.083142	1.127159	1.172887	1.196413	1.269734	1.347351	1.429502
25	1.086753	1.132795	1.180707	1.205386	1.282431	1.364192	1.450945

Day	2%	2.5%	2.75%	3%	3.5%
1	1.0000547	1.0000684	1.0000753	1.0000821	1.0000958
2	1.0001094	1.0001368	1.0001506	1.0001646	1.0001916
3	1.0001641	1.0002052	1.0022591	1.0002463	1.0002874
4	1.0002188	1.0002736	1.0003012	1.0003284	1.0003832
5	1.0002735	1.0003420	1.0003765	1.0004105	1.0004790
6	1.0003282	1.0004104	1.0004518	1.0004926	1.0005749
7	1.0003829	1.0004788	1.0005272	1.0005748	1.0006707
8	1.0004376	1.0005473	1.0006025	1.0006569	1.0007666
9	1.0004924	1.0006157	1.0006779	1.0007391	1.0008625
10	1.0005471	1.0006842	1.0007532	1.0008212	1.0009584
11	1.0006018	1.0007526	1.0008286	1.0009034	1.0010543
12	1.0006565	1.0008211	1.0009039	1.0009856	1.0011502
13	1.0007113	1.0008895	1.0009793	1.0010678	1.0012461
14	1.0007660	1.0009580	1.0010547	1.0011500	1.0013420
15	1.0008208	1.0010264	1.0011300	1.0012322	1.0014379
16	1.0008755	1.0010949	1.0012054	1.0013144	1.0015388
17	1.0009302	1.0011634	1.0012808	1.0013966	1.0016298
18	1.0009850	1.0012319	1.0013562	1.0014788	1.0017257
19	1.0010397	1.0013003	1.0014316	1.0015610	1.0018217
20	1.0010945	1.0013688	1.0015070	1.0016432	1.0019177
21	1.0011493	1.0014373	1.0015824	1.0017255	1.0020137
22	1.0012040	1.0015058	1.0016579	1.0018077	1.0021097
23	1.0012588	1.0015743	1.0017333	1.0018899	1.0220571
24	1.0013136	1.0016428	1.0018087	1.0019722	1.0023017
25	1.0013683	1.0017113	1.0018841	1.0020545	1.0023977
26	1.0014231	1.0017799	1.0019596	1.0021367	1.0024937
27	1.0014779	1.0018484	1.0020350	1.0022190	1.0025898
28	1.0015327	1.0019169	1.0021105	1.0023013	1.0026858
29	1.0015874	1.0019854	1.0021859	1.0023836	1.0027819
30	1.0016422	1.0020540	1.0022614	1.0024659	1.0028779
31	1.0016970	1.0021225	1.0023369	1.0025482	1.0029740

Index